ATOMS

LES ATOMES

PAR

JEAN PERRIN

Professeur de chimie physique à la Faculté des Sciences de Paris.

Avec 13 figures.

LIBRAIRIE FÉLIX ALCAN
1913

ATOMS

BY
JEAN PERRIN

TRANSLATED BY
D. LL. HAMMICK, M.A.

OX BOW PRESS
WOODBRIDGE, CONNECTICUT

1990 reprint by:
Ox Bow Press
P.O. Box 4045
Woodbridge, CT 06525

Library of Congress Cataloging-in-Publication Data
Perrin, Jean, 1870–1942.
   [Atomes. English]
   Atoms / Jean Perrin ; authorised translation by D. Ll. Hammick.
     p. cm.
   Translation of: Les atomes.
   ISBN 0-918024-78-1 (alk. paper). — ISBN 0-918024-79-X (pbk. :
alk. paper)
   1. Atoms. 2. Atomic theory. 3. Brownian movements. I. Title.
QC173.P42   1990
539.7—dc20   90-42918
                                                                                                     CIP

The paper used in this book meets the guidelines for permanence
and durability of the Committee on Production Guidelines for
Book Longevity of the Council on Library Resources.

Printed in the United States of America

PREFACE

Two kinds of intellectual activity, both equally instinctive, have played a prominent part in the progress of physical science.

One is already developed in a child who, while holding an object, knows what will happen if he relinquishes his grasp. He may possibly never have had hold of the particular object before, but he nevertheless recognises something in common between the muscular sensations it calls forth and those which he has already experienced when grasping other objects that fell to the ground when his grasp was relaxed. Men like Galileo and Carnot, who possessed this *power of perceiving analogies* to an extraordinary degree, have by an analogous process built up the doctrine of energy by successive generalisations, cautious as well as bold, from experimental relationships and objective realities.

In the first place they observed, or it would perhaps be better to say that everyone has observed, that not only does an object fall if it be dropped, but that once it has reached the ground it will not rise *of itself*. We have *to pay* before a lift can be made to ascend, and the more dearly the swifter and higher it rises. Of course, the real price is not a sum of money, but the external compensation given for the work done by the lift (the fall of a mass of water, the combustion of coal, chemical change in a battery). The money is only the symbol of this compensation.

This once recognised, our attention naturally turns to the question of how small the payment can be. We know that by means of a wheel and axle we can raise 1,000 kilogrammes through 1 metre by allowing 100 kilogrammes to fall 10 metres; is it possible to devise a more economical mechanism that will allow 1,000 kilogrammes to be raised

1 metre for the same price (100 kilogrammes falling through 10 metres) ?

Galileo held that it is possible to affirm that, under certain conditions, 200 kilogrammes could be raised 1 metre without external compensation, "for nothing." Seeing that we no longer believe that this is possible, we have to recognise *equivalence between mechanisms* that bring about the elevation of one weight by the lowering of another.

In the same way, if we cool mercury from 100° C. to 0° C. by melting ice, we always find (and the general expression of this fact is the basis of the whole of calorimetry) that 42 grammes of ice are melted for every kilogramme of mercury cooled, whether we work by direct contact, radiation, or any other method (provided always that we end with melted ice and mercury cooled from 100° C. to 0° C.). Even more interesting are those experiments in which, through the intermediary of friction, a heating effect is produced by the falling of weights (Joule). However widely we vary the mechanism through which we connect the two phenomena, we always find one large calory of heat produced for the fall of 428 kilogrammes through 1 metre.

Step by step, in this way the First Principle of Thermodynamics has been established. It may, in my opinion, be enunciated as follows :

If by means of a certain mechanism we are able to connect two phenomena in such a way that each may accurately compensate the other, then it can never happen, however the mechanism employed is varied, that we could obtain, as the external effect of one of the phenomena, first the other and then another phenomenon in addition, which would represent a gain.[1]

Without going so fully into detail, we may notice another similar result, established by Sadi Carnot, who, grasping the essential characteristic common to all heat engines, showed that the production of work is always accompanied " by the passage of caloric from a body at a higher tem-

---

[1] At least, the other phenomenon could only be one of those which we know can occur without external compensation (such as isothermal change of volume of a gaseous mass, according to a law discovered by Joule). In that case the gain may still be looked upon as non-existent.

perature to another at a lower temperature." As we know, proper analysis of this statement leads to the Second Law of Thermodynamics.

Each of these principles has been reached by noting analogies and generalising the results of experience, and our lines of reasoning and statements of results have related only to objects that can be observed and to experiments that can be performed. Ostwald could therefore justly say that in the doctrine of energy there are no *hypotheses*. Certainly when a new machine is invented we at once assert that it cannot create work ; but we can at once verify our statement, and we cannot call an assertion a hypothesis if, as soon as it is made, it can be checked by experiment.

Now, there are cases where hypothesis is, on the contrary, both necessary and fruitful. In studying a machine, we do not confine ourselves only to the consideration of its visible parts, which have objective reality for us only as far as we can dismount the machine. We certainly observe these visible pieces as closely as we can, but at the same time we seek to divine the *hidden* gears and parts that explain its apparent motions.

To divine in this way the existence and properties of objects that still lie outside our ken, *to explain the complications of the visible in terms of invisible simplicity*, is the function of the intuitive intelligence which, thanks to men such as Dalton and Boltzmann, has given us the doctrine of Atoms. This book aims at giving an exposition of that doctrine.

The use of the intuitive method has not, of course, been used only in the study of atoms, any more than the inductive method has found its sole application in energetics. A time may perhaps come when atoms, directly perceptible at last, will be as easy to observe as are microbes to-day. The true spirit of the atomists will then be found in those who have inherited the power to divine another universal structure lying hidden behind a vaster experimental reality than ours.

I shall not attempt, as too many have done, to decide between the merits of the two methods of research. Certainly during recent years intuition has gone ahead of induction in rejuvenating the doctrine of energy by the incorporation of statistical results borrowed from the atomists. But its greater fruitfulness may well be transient, and I can see no reason to doubt the possibility of further discovery that will dispense with the necessity of employing any unverifiable hypothesis.

Although perhaps without any logical necessity for so doing, induction and intuition have both up to the present made use of two ideas that were familiar to the Greek philosophers ; these are the conceptions of *fullness* (or continuity) and of *emptiness* (or discontinuity).

Even more for the benefit of the reader who has just read this book than for him who is about to do so, I wish to offer a few remarks designed to give objective justification for certain logical exigencies of the mathematicians.

It is well known that before giving accurate definitions we show beginners that they already possess the idea of continuity. We draw a well-defined curve for them and say to them, holding a ruler against the curve, " You see that there is a tangent at every point." Or again, in order to impart the more abstract notion of the true velocity of a moving object at a point in its trajectory, we say, " You see, of course, that the mean velocity between two neighbouring points on this trajectory does not vary appreciably as these points approach infinitely near to each other." And many minds, perceiving that for certain familiar motions this appears true enough, do not see that there are considerable difficulties in this view.

To mathematicians, however, the lack of rigour in these so-called geometrical considerations is quite apparent, and they are well aware of the childishness of trying to show, by drawing curves, for instance, that every continuous function has a derivative. Though derived functions are the simplest and the easiest to deal with, they are nevertheless exceptional ; to use geometrical language, curves that have no tangents are the rule, and regular curves, such as the circle, are interesting though quite special cases.

At first sight the consideration of such cases seems merely an intellectual exercise, certainly ingenious but artificial and sterile in application, the desire for absolute accuracy carried to a ridiculous pitch. And often those who hear of curves without tangents, or underived functions, think at first that Nature presents no such complications, nor even offers any suggestion of them.

The contrary, however, is true, and the logic of the mathematicians has kept them nearer to reality than the practical representations employed by physicists. This may be illustrated by considering, in the absence of any preconceived opinion, certain entirely experimental data.

The study of *colloids* provides an abundance of such data. Consider, for instance, one of the white flakes that are obtained by salting a soap solution. At a distance its contour may appear sharply defined, but as soon as we draw nearer its sharpness disappears. The eye no longer succeeds in drawing a tangent at any point on it ; a line that at first sight would seem to be satisfactory, appears on closer scrutiny to be perpendicular or oblique to the contour. The use of magnifying glass or microscope leaves us just as uncertain, for every time we increase the magnification we find fresh irregularities appearing, and we never succeed in getting a sharp, smooth impression, such as that given, for example, by a steel ball. So that if we were to take a steel ball as giving a useful illustration of classical continuity, our flake could just as logically be used to suggest the more general notion of a continuous underived function.

We must bear in mind that the uncertainty as to the position of the tangent plane at a point on the contour is by no means of the same order as the uncertainty involved, according to the scale of the map used, in fixing a tangent at a point on the coast line of Brittany. The tangent would be different according to the scale, but a tangent could always be found, for a map is a conventional diagram in which, by construction, every line has a tangent. An essential characteristic of our flake (and, indeed, of the coast line also when, instead of studying it on a map, we observe the line itself at various distances from it) is, on the contrary, that on any scale we *suspect*, without seeing them clearly, details that absolutely prohibit the fixing of a tangent.

We are still in the realm of experimental reality when, under the microscope, we observe the Brownian movement agitating each small particle suspended in a fluid. In order to be able to fix a tangent to the trajectory of such a particle, we should expect to be able to establish, within at least approximate limits, the direction of the straight line joining the positions occupied by a particle at two very close successive instants. Now, no matter how many experiments are made, that direction is found to vary absolutely irregularly as the time between the two instants is decreased. An unprejudiced observer would therefore come to the conclusion that he was dealing with an underived function, instead of a curve to which a tangent could be drawn.

I have spoken first of curves and outlines because curves are ordinarily used to suggest the notion of continuity and to represent it. But it is just as logical, and in physics it is more usual, to inquire into the variation of some property, such as density or colour, from one point in a given material to another. And here again complications of the same kind as those mentioned above will appear.

The classical idea is quite definitely that it is possible to decompose any material object into practically identical small parts. In other words, it is assumed that the *differentiation* of the matter enclosed by a given contour becomes less and less as the contour contracts more and more.

Now I may almost go so far as to say that, far from being suggested by experience, this conception but rarely corresponds with it. My eye seeks in vain for a small " practically homogeneous " region on my hand, on the table at which I am writing, on the trees or in the soil that I can see from my window. And if, taking a not too difficult case, I select a somewhat more homogeneous region, on a tree trunk for instance, I have only to go close to it to distinguish details on the rough bark, which until then had only been suspected, and to be led to suspect the existence of others. Having reached the limits of unaided vision, magnifying glass and microscope may be used to show each successive part chosen at a progressively increasing magnification. Fresh details will be revealed at each stage, and when at last the utmost limit of magnifying power has been reached the impression left on the mind will be very different from the one originally received. In fact, as is well known, a living cell is far from homogeneous, and within it we are able to recognise the existence of a complex organisation of fine threads and granules immersed in an irregular plasma, where we can only guess at things that the eye tires itself in vain in seeking to characterise with precision. Thus the portion of matter that to begin with we had expected to find almost homogeneous appears to be indefinitely diverse, and we have absolutely no right to assume that on going far enough we should ultimately reach " homogeneity," *or even matter having properties that vary regularly from point to point.*

It is not living matter only that shows itself to be indefinitely sponge-like and differentiated. Charcoal obtained by calcining the bark of the tree mentioned above displays the same unlimited porosity. The soil and most rocks do not appear to be easily decomposable into small homogeneous parts. Indeed, the only examples of regularly continuous materials to be found are crystals such as diamonds, liquids such as water, and gases. Thus the notion of continuity is the result of an arbitrary limitation of our attention to a part only of the data of experience.

It must be borne in mind that although closer observation of the object we are studying generally leads to the discovery of a highly irregular structure, we can with advantage often represent its properties approximately by continuous functions. More simply, although wood may be indefinitely porous, it is useful to speak of the surface of a beam that we wish to paint, or of the volume displaced by a float. In other words, at certain magnifications and for certain methods of investigation phenomena may be represented by regular continuous functions, somewhat in the same way that a sheet of tin-foil may be wrapped round a sponge without it following accurately the latter's complicated contour.

If then we refuse to limit our considerations to the part of the universe we actually see, and if we attribute to matter the *infinitely* granular structure that is suggested by the results obtained by the use of the conception of atoms, our power to apply *rigorously* mathematical continuity to reality will be found to suffer a very remarkable diminution.

Let us consider, for instance, the way in which we define the density of a compressible fluid (air, for example) at a given point and at a given moment. We picture a sphere of volume v having its centre at that point and including at the given moment a mass m. The quotient $\frac{m}{v}$ is the mean density within the sphere, and by *true* density we mean the limiting value of this quotient. This means that at the given moment the mean density within the sphere is practically constant below a certain value for the volume. Indeed, this mean density, which may possibly be notably different for spheres containing 1,000 cubic metres and 1 cubic centimetre respectively, only varies by 1 part in 1,000,000 on passing from 1 cubic centimetre to one-thousandth of a cubic millimetre. Nevertheless, even between these volume limits (the width of which is considerably influenced by the state of agitation of the fluid) variations of the order of 1 part in 1,000,000,000 occur irregularly.

Suppose the volume to become continually smaller. Instead of these fluctuations becoming less and less important, they come to be more and more considerable and irregular. For dimensions at which the Brownian movement shows great activity, for one-tenth of a cubic micron, say, they begin (in air) to attain to 1 part in 1,000, and they become of the order of 1 part in 5 when the radius of the hypothetical spherule becomes of the order of a hundredth of a micron.

One step further and the radius becomes of the same order as the molecular radius. Then, as a general rule (in a gas at any rate), our spherule will lie in intermolecular space, where its mean density will henceforth be *nil* ; at our given point the *true* density will be *nil* also. But about once in a thousand times that point will lie within a molecule and the mean density will then come to be comparable with that of water, or a thousand times higher than the value we usually take to be the true density of the gas.

Let our spherule grow steadily smaller. Soon, except under exceptional circumstances, which have few chances of occurring, it will become empty and remain so henceforth owing to the emptiness of intra-atomic space ; the true density at any given point will still remain *nil*. If, however, as will happen only about once in a million times, the given point lies within a corpuscle or the central atomic nucleus, the mean density will rise enormously and will become several million times greater than that of water.

If the spherule were to become still smaller, it may be that we should attain a measure of continuity, until we reached a new order of smallness ; but more probably (especially in the atomic nucleus, which radioactivity shows to possess an extremely complicated structure) the mean density would soon fall to nothing and remain there, as will the true density also, except in certain very rare positions, where it will reach values enormously greater than any before.

In short, the doctrine of atoms leads to the following :— density is everywhere *nil*, except at an infinite number of isolated points, where it reaches an infinite value.[1]

Analogous considerations are applicable to all properties that, on our scale, appear to be continuous and regular, such as velocity, pressure or temperature. We find them growing more and more irregular as we increase the magnification of the ever imperfect image of the universe that we construct for ourselves. Density we have seen to be *nil* at

[1] I have simplified the problem. As a matter of fact, time is a factor, and mean density, defined in a small volume v surrounding the given point *at a given instant*, must be connected with a small lapse of time τ that includes the given instant. The *mean* mass in the volume v during the time τ would be of the form $\frac{1}{\tau}\int_0^\tau m \cdot dt$, and the mean density is a second derivative with respect to volume and time. Its representation as a function of two variables would lead to infinitely indented surfaces.

all points, with certain exceptions; more generally, the function that represents any physical property we consider (say electric potential) will form in intermaterial space a *continuum* that presents an infinite number of singular points and which we shall be able to study with the aid of the mathematician.[1]

An infinitely discontinuous matter, a continuous ether studded with minute stars, is the picture presented by the universe, if we remember, with J. H. Rosny, sen., that *no* formula, however comprehensive, can embrace Diversity that has no limits, and that all formulæ lose their significance when we make any considerable departure from the conditions under which we acquire our knowledge.

The conclusion we have just reached by considering a continuously diminishing centre can also be arrived at by imagining a continually enlarging sphere, that successively embraces planets, solar system, stars, and nebulæ. Thus we find ourselves face to face with the now familiar conception developed by Pascal when he showed that man lies " suspended between two infinities."

Among those whose genius has thus been able to contemplate Nature in her full high majesty, I have chosen one, to whom I dedicate this work, in homage to a departed friend; to him I owe the inspiration that brings to scientific research a tempered enthusiasm, a tireless energy and a love of beauty.[2]

---

[1] Those who are interested in this question will do well to read the works of M. Emile Borel, particularly the very fine lecture on " Molecular Theories and Mathematics " (Inauguration of the University of Houston, and *Revue générale des Sciences*, November, 1912), wherein he shows how the physics of discontinuity may possibly transform the mathematical analysis created originally to meet the needs of the physics of continuity.

[2] Professor Perrin's book is dedicated to the memory of M. Noel Bernard.—[TR.]

CONTENTS

CHAP.		PAGE
	PREFACE	V
I.	CHEMISTRY AND THE ATOMIC THEORY	1
II.	MOLECULAR AGITATION	53
III.	THE BROWNIAN MOVEMENT—EMULSIONS	83
IV.	THE LAWS OF THE BROWNIAN MOVEMENT	109
V.	FLUCTUATIONS	134
VI.	LIGHT AND QUANTA	145
VII.	THE ATOM OF ELECTRICITY	173
VIII.	THE GENESIS AND DESTRUCTION OF ATOMS	195
	APPENDIX, 1921	218
	INDEX	229

ATOMS

CHAPTER I

CHEMISTRY AND THE ATOMIC THEORY

Molecules.

Some twenty-five centuries ago, before the close of the lyric period in Greek history, certain philosophers on the shores of the Mediterranean were already teaching that changeful matter is made up of indestructible particles in constant motion; atoms which chance or destiny has grouped in the course of ages into the forms or substances with which we are familiar. But we know next to nothing of these early theories, of the works of Moschus, of Democritus of Abdera, or of his friend Leucippus. No fragments remain that might enable us to judge of what in their work was of scientific value. And in the beautiful poem, of a much later date, wherein Lucretius expounds the teachings of Epicurus, we find nothing that enables us to grasp what facts or what theories guided Greek thought.

1.—Persistence of the Component Substances in Mixtures.—Without raising the question as to whether our present views actually originated in this way, we may notice that it is possible to infer a discontinuous structure for certain substances which, like water, appear perfectly homogeneous, merely from a consideration of the familiar properties of solution. It is universally admitted, for instance, that when sugar is dissolved in water the sugar and the water both exist in the solution, although we cannot distinguish the different components from each other. Similarly, if we drop a little bromine into chloroform, the bromine and the chloroform constituents in the homogeneous liquid thus obtained will continue to be *recognisable* by their colour and smell.

This would be easily explicable if the different substances existed in the liquid in the same way that the particles of a well powdered mixture exist side by side ; though we may no longer be able to distinguish the particles from each other even at close quarters, we can nevertheless detect them (by their colour or taste, for example, as may readily be verified by making an intimate mixture of powdered sugar and flowers of sulphur). Similarly, the persistence of the properties of bromine and of chloroform in the liquid obtained by mixing these substances is perhaps due to the existence in the liquid of small particles, in simple juxtaposition (but unmodified), which by themselves constitute bromine, and of other particles which, by themselves, form chloroform. These elementary particles, or *molecules*, should be found in all mixtures in which we recognise bromine or chloroform, and their extreme minuteness alone prevents us from perceiving them as individuals. Moreover, since bromine (or chloroform) is a *pure substance*, in the sense that no single observation has ever led us to recognise in it the properties of components of which it could be a mixture, we must suppose that its molecules are composed of the same substance.

But they may be of various dimensions, like the particles which make up powdered sugar or flowers of sulphur ; they may even be extremely minute droplets, capable, under certain circumstances, of uniting among themselves or of subdividing without losing their nature. Indefiniteness of this kind is often met with in physics when we come to give a precise meaning to a hypothesis put forward vaguely in the first place. In such circumstances we trace out as far as possible the consequences of each particular precise form of our hypothesis that we can devise. The necessary condition that they must agree with experiment or simply their obvious barrenness soon leads us to abandon most of these tentative forms and they are consequently omitted from subsequent discussion.

2.—EACH CHEMICAL SPECIES IS COMPOSED OF CLEARLY CHARACTERISED MOLECULES.—In the present case one only of the precise forms which we have been able to devise for our general hypothesis has proved fruitful. It has been assumed that the molecules that make up a pure substance are exactly identical and remain identical in all mixtures in which that substance is found. In liquid bromine, in bromine vapour, in a solution of bromine, at all pressures and temperatures, as long as we can " recognise bromine " this material " bromine " is resolvable, at a sufficient magnification, into identical molecules. Even in the solid state these molecules exist, as assemblages of objects each maintaining their own individuality and separable without rupture, (quite unlike the way in which bricks are cemented into a wall ; for when the wall is destroyed, the bricks cannot be recovered intact, whereas on melting or vapourising a solid the molecules are recoverable with their independence and mobility unimpaired).

If every pure substance is necessarily made up of a particular kind of molecule, it does not follow that, conversely, with each kind of molecule we should be able to make up a pure substance without admixture of other kinds of molecules. We can thus understand the properties of that singular gas nitrogen peroxide ; it does not obey Boyle's law, and its red colour becomes more intense when it is allowed to expand into a larger volume. These abnormalities are explained in every detail if nitrogen peroxide is really a mixture in variable proportions of two gases, the one red and the other colourless. It is certainly to be expected that each of these gases would be composed of a definite kind of molecule ; but as a matter of fact it is not possible to separate these two kinds of molecule or, in other words, to prepare in a pure state the red and the colourless gases. As soon as one brings about a separation that, for example, increases for a moment the proportion of red gas, a fresh quantity of colourless gas is at once re-formed at the expense of the red, until the proportion fixed by the particular pressure and temperature is reached once more.[1]

---

[1] A more complete discussion leads us to regard the colourless gas molecule as formed by the union of two molecules of red gas, the two gases having the chemical formulæ (in the sense that will be explained later), N_2O_4 and NO_2.

More generally, it appears that a substance may be easy to characterise and to recognise as a constituent of various mixtures, though at the same time we may not know how to separate it in the pure state from substances that dissolve it or with which it is in equilibrium. Chemists do not hesitate to speak of sulphurous acid or carbonic acid, although it is not possible to separate these hydrogen compounds from their solutions. On our hypothesis a particular kind of molecule should correspond to each *chemical species* that is to be regarded in this manner as existing ; and, conversely, to each kind of molecule will correspond a chemical entity, definite though not always capable of being isolated. Of course, we do not assume that the molecules that make up a chemical entity are indivisible like "atoms"; on the contrary, we are generally led to the view that they are divisible. But in that case the properties by means of which the chemical entity is recognised disappear and others make their appearance ; the new properties belong to new chemical entities, which have for molecules the fragments of the old molecules.[1]

In short we suppose that any substance whatever, that appears to be homogeneous to observations on our scale of dimensions, would be resolved at a sufficient magnification into well defined molecules of as many different kinds as there are constituents recognisable from the properties of the given substance.

We shall see that these molecules do not remain at rest.

3.—MOLECULAR AGITATION IS MADE MANIFEST BY THE PHENOMENA OF DIFFUSION.—When a layer of alcohol is superposed upon a layer of water, although the alcohol is on top it is well known that the two liquids do not remain separated, in spite of the fact that the lower layer is the denser. Reciprocal solution takes place, by the *diffusion* of the two substances into each other, and in a few days renders the liquid uniform throughout. It must therefore be assumed that the molecules of alcohol and of water are endowed with movement, at least during the time the act of solution lasts.

[1] A simple example is furnished in sal ammoniac, which has a molecule capable of splitting up into two portions—*i.e.*, an ammonia molecule and a hydrochloric acid molecule.

As a matter of fact, if we had superposed water and ether, a distinct surface of separation would have persisted. But even in this case of incomplete solubility, water passes into every layer of the upper liquid and ether penetrates equally into each layer of the lower liquid. A movement of the molecules is thus again manifest.

With gaseous layers, diffusion, which is more rapid, always proceeds until the entire mass becomes uniform. In Berthollet's famous experiment a globe containing carbon dioxide was put in communication by means of a stop-cock with another globe containing hydrogen at the same pressure, the hydrogen being above the carbon dioxide. In spite of the great difference in density between the two gases, the composition gradually became uniform in the two globes and soon each one of them contained as much hydrogen as carbon dioxide. The experiment led to the same result no matter what pairs of gases were used.

Moreover, the rate of diffusion has no connection with any difference in properties of the two fluids put in contact. It may be great or small for very similar bodies as well as for those that are very dissimilar. We find, for example, that ethyl alcohol (spirits of wine) and methyl alcohol (wood spirit), which chemically and physically are very similar, do not interpenetrate more quickly than ethyl alcohol and toluene, which differ much more widely from each other.

Now, if diffusion takes place between two layers of fluids of any kind—between, for instance, ethyl alcohol and water, ethyl alcohol and methyl alcohol, ethyl alcohol and propyl alcohol—may we not assume that diffusion takes place in just the same way between ethyl alcohol and ethyl alcohol ? In the light of the preceding considerations, it seems difficult to avoid the conclusion that diffusion probably does take place but that we are no longer able to perceive it on account of the identical nature of the two interpenetrating bodies.

We are thus forced to imagine a continual diffusion taking place between any two contiguous sections of the same fluid. If molecules do exist, it comes to the same thing if we say that every surface traced in a fluid is traversed incessantly by molecules passing from one side to the other, and hence *that the molecules of any fluid whatever are in constant motion.*

If these conclusions are well founded, our ideas on fluids " in equilibrium " must undergo a very profound readjustment. Like homogeneity, equilibrium is only apparent and disappears when we change the " magnification " under which we observe matter. More exactly, such equilibrium represents a particular permanent condition of unco-ordinated agitation. From our observations on the ordinary dimensional scale we can get no inkling of the internal agitation of fluids, because each small element of volume at each instant gains as many molecules as it loses and preserves the same mean condition of unco-ordinated movement. As we proceed we shall find that these ideas will become more precise, and we shall come to understand better the important position the theories of statistics and probability must occupy in physics.

4.—MOLECULAR AGITATION EXPLAINS THE EXPANSIBILITY OF FLUIDS.—Having once admitted the existence of molecular agitation, we can readily understand the expansibility of fluids, or, which comes to the same thing, why they always exert a pressure on the walls of the vessels that contain them. This pressure is due, not to a mutual repulsion between diverse portions of the fluid, but to the incessant impact of the molecules of the fluid against the walls.

This somewhat vague hypothesis was given a precise form and developed towards the middle of the eighteenth century for the case of a fluid rarefied sufficiently to possess the properties characteristic of the gaseous state. It is assumed that under such conditions the molecules roughly correspond to elastic spheres having a total volume very small compared with the space they traverse, and which are on the average so far from each other that each moves in a straight line for the greater part of its path, until impact with another molecule abruptly changes its direction. We shall see later how this hypothesis explained all the known properties of gases and how by means of it other properties then unrecognised were predicted.

Suppose that a gaseous mass is heated at constant volume ;

CHEMISTRY AND THE ATOMIC THEORY 7

we know that its pressure then rises. If this pressure is due to the impacts of the molecules upon the containing walls, we must suppose that the molecules are now moving with speeds that, on the average, have increased, so that each square centimetre of the containing surface is subjected to impacts that are more violent and more numerous. *Molecular agitation must therefore increase with rise in temperature.* If, on the other hand, the temperature falls, the molecular agitation will slacken and should tend towards zero along with the pressure of the gas. At the " absolute zero " of temperature the molecules should be absolutely motionless.

In this connection we may remember that in all cases, without exception, rate of diffusion becomes slower the lower the temperature. Thus molecular agitation and temperature always vary in the same sense and appear to be fundamentally connected with each other.

ATOMS.

5.—SIMPLE SUBSTANCES.—Amid the vast aggregate of known substances (which are in general mixtures in varying proportions) the various chemical species serve as co-ordinating centres in the same way that the four apices of a tetrahedron act as points of reference for all points inside it. But even then their number is enormous. As we know, since Lavoisier's time the study and classification of all such species has been simplified by the discovery of " simple substances," indestructible substances obtained by pushing as far as possible the " decomposition " of the different available materials.

The meaning of this word "decomposition" will be made clear by the discussion of some particular case. It is possible, for instance, by merely heating, to transform sal ammoniac, a well-defined pure solid substance, into a mixture of gases that can be separated by a suitable fractionation (diffusion or effusion) into ammonia gas and hydrochloric acid gas. Ammonia gas is in its turn transformable (by means of a stream of sparks) into a gaseous mixture of nitrogen and hydrogen, which in their turn are easily separable. Then, having dissolved the hydrochloric acid gas in a little water, it is possible (by electrolysis) to recover, firstly, the added water, and, secondly, chlorine and hydrogen (which separate at the electrodes) from the hydrochloric acid gas. From 100 grammes of the salt we can produce 26·16 grammes of nitrogen, 7·50 grammes of hydrogen, and 66·34 grammes of chlorine, these masses being equal to that of the salt that has disappeared.

All other ways of decomposing sal ammoniac, pushed to their utmost limit, are always found to end with the production of these three elementary bodies in exactly the same proportions. Speaking more generally, an enormous number of decompositions has led to the recognition of about 100 simple substances (nitrogen, chlorine, hydrogen, carbon, etc., etc.) possessing the following property:—

Any material system whatsoever can be decomposed into masses each composed of one of these simple substances; these masses are absolutely independent, in quantity and in their nature, of the operations that the given system has been made to undergo.

Thus, if we start with fixed masses of these different simple substances, we can, after making them react with each other in every conceivable way, always recover the mass of each simple substance originally taken. If the element oxygen is represented to begin with by 16 grammes, it is not within our power to bring about an operation at the end of which we do not regain 16 grammes of oxygen, neither more nor less, on decomposing the system obtained.[1]

It is therefore hard to avoid the conclusion that the oxygen has actually persisted throughout the series of compounds produced, disguised but certainly present; one and the same "*elementary substance*" must exist in all substances containing oxygen, such as water, oxygen, ozone, carbon dioxide, or sugar.[2]

[1] It is of course understood that oxygen and ozone, which are transformable in their entirety into each other, are regarded as equivalent. Similarly with all simple substances capable of existing in various allotropic modifications.

[2] Incidentally, it is not quite correct to speak of this particular elementary substance as "oxygen." Clearly, we might just as well call it "ozone," since oxygen and ozone can be completely transformed into each other. One and the same substance, to which a distinct name should be given, appears to us, according to circumstances, sometimes in the form "oxygen" and sometimes in the form "ozone." We shall perceive the significance of this later on (para. 7).

CHEMISTRY AND THE ATOMIC THEORY 9

But if sugar, for instance, is made up of identical molecules, oxygen, with its usual properties masked, must have a place in the structure of each, and similarly with carbon and hydrogen, which are the other elements in sugar. We shall endeavour to make out in what form the elementary substances exist in molecules.

6.—THE LAW OF CHEMICAL DISCONTINUITY.—Certain fundamental chemical laws will help us in this task. We have first that the proportion of an element that enters into a molecule cannot have all possible values. When carbon burns in oxygen, it produces a pure substance (carbon dioxide), containing 3 grammes of carbon to every 8 grammes of oxygen. It would not be irrational to expect (and indeed eminent chemists have in the past regarded it as possible) that, by changing the conditions under which combination takes place (by working, for example, under high pressures or by substituting slow for rapid combustion), we might be able to change slightly the proportions of combined carbon and oxygen. Thus we might not unreasonably expect to be able to obtain a pure substance possessing properties approximating to those of carbon dioxide and containing, for example, for 3 grammes of carbon, 8 grammes plus 1 decigramme of oxygen. No such substance is produced, and the fact that the absence of such substances is general gives us the "*Law of Definite Proportions*" (mainly due to Proust's work), which may be stated as follows:—

The proportions in which two elements combine cannot vary continuously.

This is not meant to imply that carbon and oxygen can unite in one single proportion only; it is not difficult (as in the preparation of carbon monoxide) to combine 3 grammes of carbon, not with 8, but with 4 grammes of oxygen. Only the variation, as we see, is in this case very large; it is, in fact, a discontinuous leap. At the same time the properties of the compound thus obtained have become very different from those of carbon dioxide. The two compounds are marked off from each other, as it were, by an insuperable gap.

The above example immediately suggests another law,

discovered by Dalton. It might be merely fortuitous that 3 grammes of carbon should unite with either 4 grammes of oxygen or with exactly double that amount. But we find simple ratios figuring in so large a number of cases that we cannot regard them as so many accidental coincidences. And this leads us to the Law of Multiple Proportions, which we can enunciate as follows :—

If two definite compounds are taken at random from among the multitude of those containing the simple substances A and B, and if the masses of the element B that are found to be combined with the same mass of the element A are compared, it is found that those masses are usually in a very simple ratio to each other. In certain cases they may be, and in fact frequently are, exactly equal.

Thus the ratio of chlorine to silver in silver chloride and in silver chlorate is found to be the same, or at least the error does not exceed that conditioned by the degree of accuracy reached in the operations of analytical chemistry. Now analytical accuracy has been increasing continuously, and in this particular case (of Stas' measurements) exceeds 1 part in 10,000, so that we cannot possibly doubt that rigorous equality holds.

7.—THE ATOMIC HYPOTHESIS.—We owe to Dalton the happy inspiration that, embracing in the simplest manner both Proust's law and the law he had discovered himself, finally gave capital importance to molecular theories in the co-ordination and prediction of chemical phenomena (1808).

Dalton supposed that each of the *elementary substances* of which all the various kinds of materials are composed is made up of a *fixed species* of particles, all absolutely identical;[1] these particles pass, without ever becoming subdivided, through the various chemical and physical transformations that we are able to bring about, and, being *indivisible* by means of such changes, they can therefore be called *atoms*, in the etymological sense.

[1] Identical when once isolated, even if they are not absolutely interchangeable at any given moment. Two springs when compressed to different extents may be regarded as identical if, when released, they become identical. There will thus be no difference between an iron atom extracted from ferrous chloride and one obtained from ferric chloride.

CHEMISTRY AND THE ATOMIC THEORY

Any single molecule necessarily contains a whole number of atoms of each elementary substance present. Its composition therefore cannot vary continuously (which is Proust's law), but only by discontinuous leaps, corresponding to the gain or loss of at least one atom (which leads us to Dalton's law of multiple proportions).

It is clear, moreover, that if a molecule were a highly complex body, containing several thousand atoms, analytical chemistry would be found to be too inexact to give us any information as to the entry or exit of a few atoms more or less. That the laws of discontinuity were discoverable when chemical analysis was not always reliable to within more than 10 per cent.[1] is clearly due to the fact that the molecules studied by chemists contained but few atoms.

A molecule may be monatomic (composed of single atoms); more usually it will contain several atoms. A particularly interesting case is that in which the atoms *combined* in the same molecule are of the same kind. We are then dealing with a simple substance which can nevertheless actually be regarded as a compound of a particular elementary substance with itself. We shall see that this is of frequent occurrence and that it explains certain cases of allotropy (we have already pointed out the case of oxygen and ozone).

In short, the whole material universe, in all its extraordinary complexity, may have been built up by the coming together of elementary units fashioned after a small number of types, elements of the same type being absolutely identical. It is easy to see how greatly the atomic hypothesis, if it is substantiated, will enable us to simplify our study of matter.

8.—THE RELATIVE WEIGHTS OF THE ATOMS WOULD BE KNOWN, IF IT WERE KNOWN HOW MANY OF EACH SORT THERE ARE IN THE MOLECULE.—Once the existence of atoms is

---

[1] We would expect, moreover, that very complicated molecules would be more fragile than molecules composed of few atoms and that they would therefore have fewer chances of coming under observation. We should also expect that if a molecule were very large (albumins ?) the entry or exit of a few atoms would not greatly affect its properties and, moreover, that the separation of a pure substance corresponding to such molecules would present no little difficulty, even if its isolation did not become impossible. And this would still further increase the probability that a pure substance easy to prepare would be composed of molecules containing few atoms.

assumed, the question arises as to how many atoms of each kind are to be found in the molecules of the better known substances. The solution to this problem will give us the relative weights of the molecules and the atoms.

Thus water, for example, contains 8 parts of oxygen for 1 part of hydrogen. Since the molecules of water are all identical, each therefore contains 8 parts of oxygen for 1 part of hydrogen. If now we know that each molecule contains n atoms o of oxygen and p atoms h of hydrogen, we know from that that the mass no must be 8 times the mass ph, which gives us the value $8\frac{p}{n}$ for the ratio $\frac{o}{h}$. At the same time we shall obtain the ratio, equal to $9n$, of the mass of the water molecule to the atom of hydrogen.

Knowing the atomic composition of, say, methane, we could obtain similarly the ratio $\frac{c}{h}$ of the mass of the carbon atom to that of the hydrogen atom. And so on.

It is thus obvious that it is sufficient to know the atomic composition of a small number of molecules to obtain, as we have shown, the relative weights of the different atoms (and of the molecules under consideration).

9.—PROPORTIONAL NUMBERS AND CHEMICAL FORMULÆ.—Unfortunately gravimetric analysis, which, by demonstrating chemical discontinuity, has led us to formulate the atomic hypothesis, provides no means for solving the problem that has just been propounded. To make this quite clear we may state that the laws of discontinuity are all summarised in the following (law of "proportional numbers"):—

Corresponding to the various simple substances:

Hydrogen, Oxygen, Carbon . . . .

we are able to find numbers (called proportional numbers):

H, O, C . . . .

such that the masses of the simple substances found in compounds are to one another as

pH qO rC . . . .

CHEMISTRY AND THE ATOMIC THEORY 13

p, q, r .... being whole numbers that often are quite simple.[1]

We can therefore express all that analysis can tell us about the substance under investigation by representing it by the chemical formula:—

$$H_p, O_q, C_r \ldots$$

Let us now replace any one of these terms in the series of proportional numbers, say C, by a term C', obtained by multiplying C by a simple arbitrary fraction, for instance, $\frac{2}{3}$, and let the other terms remain unchanged. The new series:

$$H, O, C' \ldots$$

is still a series of proportional numbers. For the compound that contains, for instance, pH grammes of hydrogen to qO grammes of oxygen and rC grammes of carbon also contains 2 pH grammes of hydrogen to 2 qO grammes of oxygen and 3 rC' grammes of carbon. Its formula, which was H_p, O_q, C_r, may now be written:

$$H_{2p}, O_{2q}, C'_{3r},$$

and if p, q, r are whole numbers, $2p, 2q$, and $3r$ will be whole numbers also.

Moreover, the new formula may possibly be simpler than

---

[1] It would not meet the case to state that these numbers are whole numbers. Let η and γ be the masses of hydrogen and carbon combined together in an analysed specimen of a hydrocarbon. These masses are only known to a certain degree of approximation, which depends on the accuracy of the analysis. However exact the analysis may be, and even if H and C were numbers chosen *quite at random*, there would always be whole numbers p and r which, within the limits of experimental error, satisfy the equation:—

$$\frac{\eta}{\gamma} = \frac{p}{r} \cdot \frac{H}{C}.$$

But the smaller values possible for p and q must increase as the analytical accuracy increases. If for an accuracy of within 1 per cent. we had found $\frac{2}{3}$ as a possible value for $\frac{p}{r}$, then the simplest possible value should become, say, $\frac{2027}{3041}$ when the degree of accuracy gets within one in a hundred thousand. Such, however, is not the case, and the value $\frac{2}{3}$ still holds with the latter degree of accuracy. The law is that the ratios of the whole numbers p, q, r .... have fixed values which appear to be simpler, and the more surprisingly so the higher the degree of accuracy attained.

the old. A compound having originally the empirical formula $H_3 C_2$ gets with the new proportional numbers the formula $H_6 C_6$, or, in other words, the formula HC.

Each of the two formulæ completely expresses *all* the information given by analytical chemistry. We therefore can obtain from analysis no evidence that will decide whether the atoms of carbon and hydrogen are in the ratio of C to H or of C′ to H, nor any means of estimating how many atoms of each kind a given molecule contains.

In other words :

There is a wide choice of *distinct* series of proportional numbers—distinct in the sense that they do not give the same formula to the same compound.[1] We pass from any one series to another by multiplying one or more terms by simple fractions. Neither analytical chemistry, however, nor the laws of discontinuity furnish the slightest clue to the recognition, among the possible series, of one in which the terms are in the same ratio as the masses of the atoms (supposing the latter to exist).

10.—SIMILAR COMPOUNDS.—Fortunately there are other considerations that can help us in our choice, which, from the point of view of analysis alone, must remain indeterminate. And in fact we have never seriously hesitated in our choice between more than a few lists of proportional numbers.

For from the first it has been held that analogous formulæ must be used to represent compounds that resemble each other. We find a case of this resemblance between the chlorides, bromides, and iodides of any given metal. These three salts are *isomorphous*, that is to say, they have the same *crystalline shape* [2] and may be made (by the evaporation of a mixed solution) to yield mixed crystals which still retain that shape (these mixed crystals are homogeneous solid mixtures of arbitrary composition). In addition to this physical resemblance, which in itself is sufficiently remarkable, the three salts resemble each other in their several

---

[1] Two series are not distinct if one is got by multiplying all the terms in the other by the same number.

[2] In the crystallographic sense ; it is possible to orientate two such crystals so that each facet of one is parallel to a facet on the other.

CHEMISTRY AND THE ATOMIC THEORY

chemical reactivities. The atoms of chlorine, bromine, and iodine therefore probably play very similar *rôles*, and their masses are probably in the same ratios as the masses of the three elements that combine with a given mass of the same metal. This at once enables us to eliminate from the series of proportional numbers which *a priori* might provide atomic ratios, all these in which the proportional numbers Cl, Br, I, corresponding to chlorine, bromine and iodine, are not in the ratios of 71 to 160 to 254.

The probable values for the atomic ratios of various alkali metals are readily obtainable, and this will still further reduce the number of possible series. But we shall not (or, at any rate, we have not done so up to the present) obtain by this means the ratio between the atomic masses of chlorine and potassium, since these two elements do not play analogous *rôles* in any class of compound. Nor, up to the present, have we been able to pass, by any definite isomorphic relation or chemical analogy, from one of the alkali metals to one of the other metals.

In short, the original lack of definiteness, so discouraging at the outset, is thus very considerably diminished. It has not been done away with altogether. And, to quote an example which has provided material for much lively controversy in the past, the study of isomorphism has furnished no adequate reason for giving to water the formula H_2O rather than HO ; that is to say, for assigning the value 16 to the ratio $\frac{O}{H}$ instead of 8.

11.—EQUIVALENTS.—We must remember that, for many chemists, who felt that little importance could be attached to the atomic theory, the question had no great interest. It appeared to them more dangerous than useful to employ a hypothesis deemed incapable of verification in the exposition of well-ascertained laws. They also held that there was nothing to guide their choice among the possible series of proportional numbers except the one condition that *the facts* should be expressed in language as clear as possible. The use of the hypothesis was of advantage in that the memorisation and prediction of reactions was facilitated, and the representation of similar compounds by analogous formulæ was made possible ; but, apart from this, it only remained to assign the simplest formulæ to compounds deemed the most important. For instance, it seemed reasonable to write HO for the formula of water, thus arbitrarily choosing the number 8 from the possible values for the ratio $\frac{O}{H}$.

In this way scientists hostile or indifferent to the atomic theory agreed in using a particular series of proportional numbers under the name of "*equivalents.*" This *equivalent notation*, adopted by the most influential chemists and prescribed in France in the curriculum used in elementary schools,[1] hindered the development of chemistry for more than fifty years. In fact, putting all question of theory on one side, it has shown itself very much less successful in representing and suggesting phenomena than the *atomic notation* proposed by *Gerhardt* about 1840. In this notation those proportional numbers are used that Gerhardt and his successors, for reasons which will appear presently, regarded as giving the ratios of atomic weights that isomorphism and chemical analogy had not been able to determine.

Avogadro's Hypothesis.

12.—Laws of Gaseous Expansion and Combination.— The considerations that have made these most important numbers known to us depend upon the now familiar gas laws.

To begin with, it has been known since the time of Boyle (1660) and Marriotte (1675) that at a fixed temperature the density of a gas (mass contained in unit volume) is proportional to the pressure.[2] Let, therefore, n and n' be the numbers of molecules present per cubic centimetre of two different gases at the same temperature and pressure. If we multiply the pressure common to the two gases by the

---

[1] Until about 1895.
[2] As a matter of fact, we are dealing with a law valid only within limits. It is fairly well satisfied (to within about 1 per cent.) for the various gases when their pressure is less than ten atmospheres, much better at still lower pressures, and apparently becomes rigorously exact as the density tends to zero.

CHEMISTRY AND THE ATOMIC THEORY 17

same number, say 3, the masses contained per cubic centimetre are multiplied by 3, and, consequently, the numbers n and n' also ; for a given temperature the ratio $\frac{n}{n'}$ of the numbers of molecules present per cubic centimetre in the two gases at the same pressure is independent of that pressure.

Further, Gay-Lussac showed (about 1810) that, at fixed pressure, the density of a gas varies with the temperature in a manner independent of the particular nature of the gas [1] (thus oxygen and hydrogen expand equally as their temperature is raised). Consequently in this case also, since the numbers n and n' change in the same way, their ratio does not alter.

In short, within the limits of applicability of the gas laws, whether hot or cold, under high or low pressure, the numbers of molecules present in two equal globes of oxygen and hydrogen remain in a constant ratio, provided the temperature and pressure are the same in the two globes. And similarly for all gases.

These various *fixed* ratios must be *simple*. This appears from other experiments carried out about the same time (1810) by means of which Gay-Lussac showed that :—

The volumes of gas that appear or disappear in any reaction are in simple ratios to each other.[2]

An example will make this clear. Gay-Lussac found that when hydrogen and oxygen combine together to form water, the masses of hydrogen, oxygen, and water vapour that are concerned, when reduced to the same conditions of temperature and pressure, occupy volumes that are to one another *exactly* as 2 : 1 : 2. Let n be the number of oxygen molecules per cubic centimetre and n' the number of water vapour molecules. The oxygen molecule contains a whole number, which is probably small, say p, of oxygen atoms. The water molecule similarly contains p' atoms of oxygen. If no oxygen is lost, the number np of the atoms making up the

---
[1] Here again the law is limited in its application, being better satisfied the smaller the density.
[2] As a matter of fact, Gay-Lussac did not draw from this statement the proposition in molecular theory that is indicated here.

A.   c

oxygen that disappears must equal the number $2n'p'$ present in the water that makes its appearance. The ratio $\dfrac{n}{n'}$ is therefore equal to $2\dfrac{p'}{p}$, and is moreover a simple fraction, since p and p' are small whole numbers.

But as yet we have had no indication that the matter is even simpler than might be supposed ; in other words, that the numbers n and n' must invariably be equal.

13.—AVOGADRO'S HYPOTHESIS.—The famous hypothesis of Avogadro (1811) asserts this equality. Having made the preceding observations on the subject of Gay-Lussac's laws, this chemist laid it down that *equal volumes of different gases, under the same conditions of temperature and pressure, contain equal numbers of molecules.* The hypothesis may also be enunciated with advantage as follows :—

When in the gaseous condition, equal numbers of molecules of any kind whatever, enclosed in equal volumes at the same temperature, exert the same pressure.[1]

This proposition, which was at once defended by Ampère, provides, if it be true, as we shall see later and as Ampère pointed out, " a method for determining the relative masses of the atoms and the proportions according to which they enter into combination." But Avogadro's theory, put forward as it was accompanied by other inexact considerations, and being as yet without sufficient experimental foundation, was received by chemists with great suspicion. We owe the recognition of its supreme importance to Gerhardt, who, not content with vague suggestions that had convinced nobody, proved in detail [2] the superiority of the notation which he deduced from the theory and which from his time forward has gained so many adherents that it is now accepted everywhere without opposition. An account of these past controversies would have no interest for us at this point, and we are solely concerned with understanding how Avogadro's hypothesis is able to give us the ratios of the atomic weights.

[1] It is obvious that the hypothesis, supposing that it holds good, will be the more rigorously applicable the more accurately the laws of " perfect " gases are obeyed ; that is to say, the smaller the gas density.

[2] " Précis de chimie organique."

CHEMISTRY AND THE ATOMIC THEORY 19

14.—ATOMIC COEFFICIENTS.—Let us imagine certain absolutely identical vessels, of volume V, filled with the various pure substances known in the gaseous state, at the same temperature and pressure. If Avogadro's hypothesis is correct, the gaseous masses thus obtained will contain the same number of molecules, say N, which is proportional to V.

Let us consider the hydrogen compounds in particular. In every case the molecule contains the mass h grammes of the hydrogen atom a whole number of times, say p; the corresponding containing vessel therefore contains Nph grammes of hydrogen, that is, p times H grammes, where H is the product Nh, which is independent of the given substance since N is the same for all. *Equal volumes of different hydrogen compounds therefore all contain a simple multiple of a fixed mass of hydrogen.*[1]

Similarly, for the oxygen compounds, each of our containing vessels must contain a whole number of times, say q, the mass O grammes of oxygen (which is equal to No, where o is the mass of an oxygen atom); for the carbon compounds, each vessel must contain r times (r being a whole number) the mass C grammes of carbon (equal to Nc, c being the mass of the carbon atom); and so on. Since finally the numbers H, O, C . . . . are proportional to N, we could, if desired, choose the volume V in such a way that one of these numbers, say H, has any desired value, unity, for instance. All the others will then be fixed.

These consequences of Avogadro's hypothesis have been fully confirmed by chemical analysis and the measurement of densities in the gaseous state, for *thousands of substances*, no single exception [2] having been discovered. At the same

[1] But the reverse is not necessarily the case; thus, suppose that Avogadro's hypothesis is incorrect, so that N, N', N" . . . . are the numbers of molecules in volume V of various hydrogen compounds; let p, p', p'' . . . . be the whole numbers of hydrogen atoms present in each respective gaseous molecule. To say that the masses of hydrogen $Nph, N'p'h, N''p''h$ . . . . contained in equal volumes V are simple multiples of a fixed mass H only implies that $Np, N'p', N''p''$, ... and consequently N, N', N" . . . . are to one another in simple ratios. This is the important proposition deduced in para. 12 from the law of gaseous combination (which, incidentally, is thus established on a far broader experimental basis than was available to Gay-Lussac); it is not, however, 'Avogadro's more exact theorem.

[2] We might regard bodies such as nitrogen peroxide (see para. 2), *which do not*

time, the numbers H, O, C . . . . corresponding to every value of the volume V are available.

In other words, temperature and pressure being fixed, a volume V can be found (about 22 litres under normal conditions [1]), such that those of our containing vessels that contain hydrogen will contain exactly, or almost exactly, 1 gramme (hydrochloric acid, chloroform), or almost exactly 2 grammes (*water*, acetylene, *hydrogen*), or almost exactly 3 grammes (ammonia), or almost exactly 4 grammes (methane, ethylene), or almost exactly 5 grammes (pyridine), or almost exactly 6 grammes (benzene), but never intermediate qualities, such as 1·1 or 3·4 grammes.

For the same volume each vessel will contain, if the simple substance oxygen enters into the composition of the compound enclosed therein, either exactly 16 grammes of oxygen (*water*, carbon monoxide), or exactly twice 16 grammes (carbon dioxide, *oxygen*), or exactly 3 times 16 grammes (sulphuric anhydride, *ozone*), etc. . . . . but never intermediate quantities, such as 5·19 or 3·7 grammes.

Still at the same volume, our containing vessels will contain either no carbon at all or exactly 12 grammes of the substances (methane, carbon monoxide), or exactly twice 12 grammes (acetylene), or exactly 3 times 12 grammes (acetone), etc., always without intermediate quantities.

Similarly the vessels will contain, if chlorine, bromine, or iodine exist therein, a whole number of times 35·5 grammes of chlorine, 80 grammes of bromine, and 127 grammes of iodine, so that (according to Avogadro's hypothesis) the masses of the three atoms corresponding should be to one another as 35·5 : 80 : 127. It is very remarkable that we should obtain in this way numbers in the very same ratio that was suggested by the isomorphism and chemical analogies

---

obey the laws of Boyle and Gay-Lussac, and which consequently *do not come within the scope* of the present discussion, as constituting exceptions. But we have pointed out that nitrogen peroxide does not obey the gas laws, because it is not a single gas but a mixture in varying proportions of two gases. Analogous remarks apply to certain anomalous cases that at first sight seem important (*e.g.*, sal ammoniac vapour).

[1] At the temperature of melting ice and under atmospheric pressure (76 cms. of the barometric mercury column, at Paris).

CHEMISTRY AND THE ATOMIC THEORY 21

between chlorides, bromides, and iodides (para. 10). This agreement obviously supports Avogadro's hypothesis.

Thus step by step it has been possible to obtain experimentally, from the densities of gases, a series of remarkable proportional numbers—

$$H = 1, \quad O = 16, \quad C = 12, \quad Cl = 35 \cdot 5 \ldots$$

which are in the same ratios as the atomic weights, if Avogadro's hypothesis is correct, and which are, at any rate for those among them that can be subjected to the test, quite in accordance with the ratios fixed already by the facts of isomorphism and chemical analogy.

For the sake of brevity, it has become customary to call these numbers *atomic weights*. It is more correct (since they are numbers and not weights or masses) to call them *atomic coefficients*. Moreover, it is customary to speak of the mass of a simple substance that, in grammes, is measured by its atomic coefficient as a *gramme atom* of that body: 12 grammes of carbon or 16 grammes of oxygen are the gramme atoms of carbon and oxygen.

15.—DULONG AND PETIT'S LAW.—We shall now, in order to deal with the isomorphism and analogies between simple substances that form no volatile compounds, examine more closely the atomic ratios of all the simple substances. Where some uncertainty still exists in regard to a small number of metals that show no obvious analogies to substances having atomic weights that are already known, we can remove it by the application of a rule discovered by Dulong and Petit.

According to this rule, when the specific heat of a simple substance *in the solid state* is multiplied by its atomic weight, very nearly the same number is obtained in all cases; this number is about 6. We may express this result more clearly as follows :—

In the solid state nearly the same quantity of heat is required, namely, about 6 calories, to raise the temperature of any gramme atom through 1° C.

If, therefore, there is any doubt as to the value to be assigned to an atomic coefficient—for example, to that of gold—we need only observe that the specific heat of gold is

·03 to conclude that its atomic coefficient must be in the neighbourhood of 200. It can then be accurately fixed by the chemical analysis of gold compounds ; gold chloride, for instance, contains 65·8 grammes of gold to 35·5 grammes of chlorine, so that the " atomic weight " of gold must be a simple multiple or sub-multiple of 65·3. Seeing that it must be in the neighbourhood of 200, it is therefore most probably equal to 197, which is 3 times 65·7.

It goes without saying that a determination of this kind, depending as it does upon an empirical rule, cannot be held to have the same value as those based upon isomorphism and Avogadro's hypothesis. Such a reservation is all the more necessary since certain elements (boron, carbon, silicon) do not obey Dulong and Petit's rule with certainty, at any rate at ordinary temperatures.[1] The number of such exceptions, and the seriousness of the discrepancies they show, increases moreover as the temperature falls and the specific heat ultimately tends towards zero [2] for all elements (Nernst), so that the rule becomes entirely false at low temperatures (for instance, the atomic heat for diamond below $-240°$ C. is less than ·01).

We cannot, however, regard the very numerous instances of agreement pointed out by Dulong and Petit (and afterwards by Regnault) as entirely fortuitous, and we need only modify their statement, giving it the following form, which includes all recent results :—

The quantity of heat required to raise, at constant volume,[3] the temperature of a solid mass through $1°$ C. is practically nothing at very low temperatures, but increases as the tem-

[1] The specific heat of the gramme atom is at ordinary temperatures, instead of 6, 4·5 for silicon, 3 for boron, 2 for carbon.

[2] *Cf.* Dewar (Proc. Roy. Soc., 1913, A 89) ; he has shown that the specific heat of the elements at very low temperatures is a " periodic " function of their atomic weights [TR.].

[3] The heat used up in the form of work done against the forces of cohesion can easily be calculated if the compressibility is known, and must be deducted, according to Nernst, from the gross value obtained in the usual determination of specific heat. Ultimately (*cf.* Pierre Weiss' work on ferromagnetic bodies) it would be necessary to subtract the heat required to destroy the natural magnetisation of the body. In order to obtain " absolute " results, only that portion of the heat absorbed must be taken into account that appears to be concerned in increasing the potential and kinetic energy of the various atoms, which are maintained at a constant mean distance from each other.

CHEMISTRY AND THE ATOMIC THEORY 23

perature rises, finally becoming very nearly constant.[1] It is then about 6 calories per gramme atom, independent of the nature of the atoms composing the solid mass.

This limit is reached the more rapidly with the elements of higher atomic weight; thus it is practically reached in the case of lead (Pb = 207) at about —200° C., and in the case of carbon not until above 900° C.

It is important to remember that compound substances obey the law. This is the case at ordinary temperatures for the fluorides, chlorides, bromides, and iodides of various metals, but not for oxygen compounds. A piece of quartz weighing 60 grammes, made up of 1 gramme atom of silicon and 2 of oxygen, absorbs only 10 calories per degree. But above 400° C.[2] it absorbs uniformly 18 calories per degree, which is exactly 6 for each gramme atom.

We are led to suspect that some important law lies behind the above facts; the atomic notation has brought it to our notice, but the kinetic theory alone is able to furnish an approximate explanation of it (para. 91).

16.—A CORRECTION.—We have seen that one of the atomic coefficients is arbitrarily fixed, and we have agreed that the smallest among them, that of hydrogen, should be taken as 1. This, indeed, was the convention first adopted, and it gives, as we have seen, 16 and 12 for the atomic coefficients of oxygen and carbon. But more accurate measurements subsequently showed that these values are somewhat too high by about 1 per cent. It then seemed desirable to alter the original convention and to agree to give to oxygen (which takes part more often than hydrogen in well-defined quantitative changes) the atomic coefficient 16 exactly. Hydrogen then becomes, to within 1 part in about 2,000, 1·0076 (as the mean of concordant values obtained by very different methods). Carbon remains 12·00 to within less than 1 part in 1,000.

Beyond this the preceding considerations require no further qualification, except that the volume V of our identical recep-

---

[1] Of course, if the body melts or volatilises, the above proposition no longer applies.
[2] According to Pionchon's measurements, carried out up to 1,200° C.

tacles (filled with various gaseous substances at a fixed temperature and pressure) are to be regarded as chosen so that those which contain oxygen will contain exactly 16 grammes or some multiple of 16 grammes.

17.—Prout's Hypothesis : Isotopes : Mendélejeff's Rule.—We have seen, in the preceding paragraph, that the difference between the atomic coefficients of carbon and oxygen is exactly 4, which is very nearly 4 times the coefficient of hydrogen. To account for this and other similar cases Prout supposed that the different atoms are built up by the union, without loss of weight (into extremely stable complexes, which cannot be decomposed), of a necessarily whole number of proto-atoms, *all of the same kind.* This hypothesis has had a remarkable history. At first received with favour, it was abandoned when accurate measurements proved, for example, that 35·457 is the atomic coefficient of chlorine, and 28·3 that of silicon. That the hypothesis has nevertheless some claim to be retained becomes obvious on reading through the list of atomic coefficients, of which the first twenty-one are printed below, in order of increasing magnitude (with the exception of one juxtaposition in the case of argon).

Hydrogen, H = 1·0076.

Helium, He = 4·0 ; lithium, Li = 6·95 ; glucinum, Gl = 9·1 ; boron, B = 10·9 ; carbon, C = 12·00 ; nitrogen, N = 14·01 ; oxygen, O = 16·000 ; fluorine, F = 19·0.

Neon, Ne = 20·2 ; sodium, Na = 23·00 ; magnesium, Mg = 24·3 ; aluminium, Al = 27·1 ; silicon, Si = 28·3 ; phosphorus, P = 31·0 ; Sulphur, S = 32·0 ; chlorine, Cl = 35·46.

Argon, A = 39·9 ; potassium, K = 39·1 ; calcium, Ca = 40·1 ; scandium, Sc = 44, etc.

If the atomic coefficients were distributed at random, we should expect four out of these twenty-one elements to have a whole number for coefficient to within ·1,[1] whereas,

---

[1] For the fifth part of a large number of points marked at random along a scale graduated in centimetres subdivided into millimetres fall into the sections, each 2 millimetres wide, that contain the centimetre divisions.

excluding oxygen (for which a whole number coefficient has been assumed), eleven elements are found to do so. We should expect that one element only would have a whole number coefficient to within ·02, whereas this is actually the case with nine of them.

The mystery was solved when the facts of radioactive transmutation led Soddy, and subsequently Fajans, to the view that two atoms might have different masses, their chemical properties being at the same time so similar that the ordinary methods of separation (which depend upon chemical affinity and cohesion) fail completely. Such elements are ISOTOPES (*same place* in the series of the elements). The forces of inertia, however, make the separation possible. A sufficiently energetic *centrifugal fractionation* (para. 57) should be capable of bringing it about. In a simpler fashion, but without leading to appreciable quantities of separated material, separation of two isotopes occurs when they are moving as positive rays (para. 103), in which condition their unequal inertia produces unequal deviations in electrostatic and magnetic fields. In this way Aston has been able to show that ordinary chlorine is a mixture of two isotopes having whole-number atomic coefficients of 35 and 37, and also that ordinary silicon is a mixture of two isotopes with atomic coefficients 28 and 29.

Thus Prout's hypothesis has been definitely verified and becomes a fundamental law. (See para. 103 and Appendix.)

Another very surprising regularity, pointed out by Mendélejeff, is brought out in the preceding list of coefficients, in which helium, neon, and argon (of zero valency) ; lithium, sodium, potassium (univalent alkali metals) ; glucinum, magnesium, calcium (divalent alkaline earth metals), and so on, are found in corresponding positions. We have here an indication of the following law, which, however, cannot now be discussed at any length :

When we arrange the atoms in order of ascending mass, we find a succession of atomic series that are analogous, element for element, with previous atomic series.

Properties other than mass (volume of the solid gramme-atom, characteristic X-rays, etc.) also lead to an arrangement of the elements in the same order, with the result that each element has a definite *sequence number* (*numéro d'ordre*) (for example, 6 for carbon and 17 for each of the two chlorine isotopes). (See table at end of Appendix.)

18.—GRAMME MOLECULES AND AVOGADRO'S NUMBER.— In order to arrive at the atomic coefficients we have considered certain identical receptacles, full of various substances in the gaseous state, at the same temperature and under a pressure such that there are exactly 16 grammes of oxygen, or some multiple of 16 grammes, in those receptacles containing oxygen compounds. The masses of pure substances that fill our receptacles under these conditions are often called gramme molecules.

The gramme molecules of various substances are those masses which in the rarefied gaseous state (at the same temperature and pressure) all occupy equal volumes, the common value for these volumes being fixed by the condition that, among those which contain oxygen, the ones containing the least oxygen shall contain exactly 16 *grammes of it.*

More briefly, but without bringing out the theoretical significance of the theorem, we can say :—

The gramme molecule of a body is the mass of it in the gaseous state that occupies the same volume as 32 grammes of oxygen at the same temperature and pressure (*i.e.*, very nearly 22,400 c.c. under " normal " conditions).

According to Avogadro's hypothesis, every gramme molecule should be made up of the same number of molecules. This number N is what is called *Avogadro's Constant* or *Avogadro's Number*.

Suppose that 1 gramme molecule contains 1 gramme atom of a certain element ; in other words, suppose that each of the N molecules of the gramme molecule is composed of 1 atom of that element, so that its gramme atom is made up of N atoms. The mass of each of these atoms is then obtainable by dividing the corresponding gramme atom by Avogadro's number, just as that of a molecule is obtained by dividing the corresponding gramme molecule by this number N. The mass o of the oxygen atom is $\frac{16}{N}$, the

CHEMISTRY AND THE ATOMIC THEORY 27

mass h of the hydrogen atom is $\frac{1\cdot 0076}{N}$, the mass co_2 of the carbon dioxide molecule is $\frac{44}{N}$, and so on. Having found Avogadro's number, we should be able to find the masses of all molecules and atoms.

19.—MOLECULAR FORMULÆ.—A gramme molecule containing N molecules composed of p hydrogen, q oxygen, and r carbon atoms also contains pH grammes of hydrogen, qO grammes of oxygen, and rC grammes of carbon. The formula $H_pO_qC_r$, which clearly expresses the number of atoms of each kind in the gramme molecule, is called a *molecular formula*.

The examples given above (para. 14) in demonstrating how gas densities indicate the atomic ratios show that the molecular formula of water is H_2O (and not HO), that of methane being CH_4 and that of acetylene C_2H_2. It is also evident, and moreover a point of considerable interest, that the formula H_2 must be assigned to hydrogen (which thus appears to be a diatomic compound), O_2 representing oxygen and O_3 ozone.[1] There are monatomic molecules also—such as those that make up the vapours of mercury, zinc, and cadmium.

MOLECULAR STRUCTURE.

20.—SUBSTITUTION.—The importance of the chemical notation imposed by Avogadro's hypothesis is particularly well illustrated in the power it gives us of representing and predicting chemical reactions. In particular, the idea of chemical substitution, which is so important in organic chemistry, is directly suggested by this notation.

Suppose that we mix chlorine with some methane, which has the molecular formula CH_4, and that we expose the mixture to the action (indirect) of light. The mixture undergoes a change, and soon, besides hydrochloric acid, it

---

[1] It is clearly no more logical to speak of an "atom of oxygen" than of an "atom of ozone." To each variety of atom should correspond a name distinct from the names of the various bodies that can be formed by the combination of such atoms among themselves.

will be found to contain as components [1] four substances, having the following molecular formulæ :—CH_4 (methane), CH_3Cl (methane monochloride or methyl chloride), CH_2Cl_2 (methylene dichloride), $CHCl_3$ (chloroform), and CCl_4 (carbon tetrachloride).

We pass from each formula to the next by writing Cl for an H, and the question inevitably arises whether the corresponding chemical reaction does not consist merely in the substitution of 1 atom of chlorine for 1 atom of hydrogen without further disturbance and without modification of the molecular structure. However natural such a hypothesis may seem, it is nevertheless still a hypothesis, for some alteration might certainly be expected to result in the situation and nature of the atomic unions when the grouping loses 1 atom of hydrogen and gains 1 atom of chlorine.

21.—An Attempt to Determine Atomic Weights from purely Chemical Considerations.—Some chemists have thought to find in substitution an accurate means for arriving at the ratios of the atomic weights, thus dispensing with the necessity of appealing to gas densities and Avogadro's hypothesis. It seems desirable to give some account of their line of reasoning, which, though instructive, is certainly not sufficiently rigorous.

Thus if we could, when in complete ignorance of molecular formulæ (which is the whole point), consider ourselves justified in regarding it as probable that the hydrogen in methane can be "replaced" in four stages, we could not avoid the conclusion that the methane molecule probably contains 4 atoms of hydrogen. Now this molecule, like any other mass of methane, weighs (according to gravimetric analysis) 4 times as much as the hydrogen it contains; the methane molecule therefore weighs 16 times as much as the hydrogen atom. We should find, with a like degree of probability and by similar processes, that the benzene molecule contains 6 atoms of hydrogen and weighs 78 times as much as one hydrogen atom. The molecular masses of

---

[1] Which could be separated by fractionation or simply identified in the mixture, if it is assumed that we know how to prepare by other means these same bodies in the pure state.

CHEMISTRY AND THE ATOMIC THEORY 29

methane and benzene are thus in the ratio of 16 to 78. Further, the carbon in the methane molecule (as in any mass of methane) weighs 3 times more than the hydrogen it contains, and hence is 12 times as heavy as the hydrogen atom; and this carbon probably constitutes a single atom, for no substance studied in this way, by substitution methods, ever gives a smaller ratio between the carbon contained in its molecule and the hydrogen atom. The carbon in the benzene molecule, which weighs 12 times as much as the 6 hydrogen atoms in it, that is to say 72 times as much as the hydrogen atom, is therefore made up of 6 carbon atoms.

We should thus obtain, from a purely chemical standpoint, the ratio $\frac{1}{12}$ for the atomic mass of hydrogen to that of carbon, with the molecular formulæ CH_4 and C_6H_6 for methane and benzene and the ratio $\frac{16}{78}$ between their molecular masses.

Two masses of these substances in the ratio of 16 to 78 will therefore each contain as many molecules as the other. Now density measurements show that the masses of methane and benzene which, in the gaseous state, occupy the same volume at the same temperature and under the same pressure are to each other as 16 is to 78 exactly, and should in consequence contain the same number of molecules. This result, established on a general basis, would give us Avogadro's hypothesis, but this time as a law and not as a hypothesis.

There will be no difficulty in filling in the details of this seductive theory, which has recently assumed great importance in education, mainly owing to the efforts of Lespieau and L.-J. Simon. Its value is undoubted in the sense that it is only through the consideration of the phenomena of substitution that we are able to obtain certain molecular formulæ (such as that of acetic acid, for example). I nevertheless am strongly of G. Urbain's opinion that it is not capable of providing practically and in a logical fashion the ratios of the weights of all atoms.

In the first place, I know of no case where it has actually been of use in obtaining atomic weights, all of them having been fixed already by the means summarised above. Moreover, whilst admitting that the theory might have developed independently, I very much doubt whether it would have proved convincing. Certainly, if we incautiously grant that it is proved by experiment that the hydrogen in methane can be replaced in four stages, then the rest follows. But would the word " replaced," which the molecular formulæ, *if we suppose that they are known*, at once suggest, have been suggested by chemical reactions alone and by the examination, without preconceived ideas, of the products of reaction ?

It is, of course, true that the products of the progressive action of chlorine on methane resemble each other as closely as do the various alums, for example, or the chlorides, bromides, and iodides of the same metal. In the latter case the analogy is so striking that the idea of substitution is forced upon one (although as a matter of fact the word has never been used in connection with such cases), and, indeed, it has proved, as we have seen, a most valuable guide in our choice of atomic weights, *at a time when no other guidance was available*.

On the other hand, it is doubtful whether chemists *really* ignorant of the formula of methane would have been able to recognise analogies between methane and methyl chloride complete enough to establish identity of molecular structure. They might equally well have assumed (taking one only of the possible hypotheses) that, taking the atomic weights of carbon and hydrogen as 6 and 1, the formulæ of the two substances in question are CH_2 and $CH_2 . CH_2Cl$, thus making methyl chloride an additive compound. And need it be pointed out that, for half a century, the majority of chemists, although perfectly well aware that potassium displaces a portion only of the hydrogen in the water it attacks, actually gave to water the formula HO and the formula $KO . HO$ to potassium hydroxide, thus regarding the latter substances as an additive compound, whereas we now look upon it as a substitution product of formula KOH, because we have assigned to water the formula HOH ?

CHEMISTRY AND THE ATOMIC THEORY 31

In short, a purely chemical theory that is able to yield us atomic coefficients and molecular formulæ has not yet been put forward, and it seems doubtful whether, starting with the facts actually known, it is possible to formulate one that does not tacitly assume a previous knowledge of the coefficients and of certain fundamental molecular formulæ, such as, for instance, that of water.

22.—MINIMUM INTERNAL DISLOCATION OF THE MOLECULE DURING REACTION : VALENCY.—As we have seen above, the possibilities of substitution suggested by the examination of the molecular formulæ enable us to predict and interpret an immense number of reactions and in this way provide a striking confirmation of Avogadro's hypothesis. Fresh hypotheses are necessary, however, to define and expand the conception of substitution.

When we say that methane CH_4 and methyl chloride CH_3Cl have the same molecular structure, we imply that the group CH_3 has not been modified by the chlorination and that it is connected with the Cl atom in the same way that it was with the H atom. This is a postulate constantly used in chemistry ; we argue continually (without always saying so clearly enough) as though the reacting molecule always undergoes the smallest possible internal disturbance compatible with reaction. It is assumed, for example, that the group CH_3 in methyl chloride *exists* in the molecule CH_4O of methyl alcohol (which is consequently written CH_3OH) because the action of hydrochloric acid HCl on this alcohol gives (together with water HOH) the methyl chloride CH_3Cl, with which we are already familiar.

Thus, when a structure made up of parts held together with screws and bolts is taken to pieces, it may be possible to remove and keep intact the whole of one important part and ultimately to incorporate it, making use of the same bolts or fastenings, into a second structure. This rough image makes it sufficiently clear how it is possible to have substitution, not only of one atom by another, but also of one group of atoms by another group ; and even the nature of the union devised to maintain our imaginary structure is found to correspond well enough with our ideas on chemical combination.

We have not yet put forward any suggestion as to the nature of the forces that keep the atoms grouped together within the molecule. It may be that each atom in the molecule is joined to *each* of the others by an attraction that varies according to their nature and decreases rapidly with the distance between them. But such a hypothesis leads to no verifiable conclusions and presents considerable difficulties. If all hydrogen atoms are attracted by all other hydrogen atoms, why is it that the only molecule built up of hydrogen atoms is H_2, the capacity of the hydrogen atom for combining with itself being exhausted directly two atoms become united? It appears as though each atom of hydrogen stretches out a single *hand* only. Directly this hand succeeds in gripping another hand, the capacity for combination of the atom is exhausted; the hydrogen atom is therefore said to be *monovalent* (or better, univalent).

Speaking more generally, we regard the atoms in a molecule as being held together by hooks or " hands " of some kind, each bond uniting two atoms only, *without any disturbing effect whatever on the other atoms present*. Of course, no one imagines that there actually are little hooks or hands on the atoms, though the absolutely unknown forces that unite them would seem to be equivalent to bonds of some such kind, which are called valencies to avoid the use of expressions that are too anthropomorphic.

If all atoms were monovalent a single molecule could never contain more than two atoms; there must therefore be polyvalent atoms. Since there is no limit to the number of atoms in a molecule except that set by the latter's fragility, which becomes progressively greater the more atoms there are in the molecule (whereas the number of children that can form a circle by taking hold of hands is not limited). Oxygen, for instance, is at least bivalent, since its atoms can form the ozone molecule O_3 as well as the molecule O_2.

The image we have used above serves to suggest that the number of valencies assumed by an atom may vary

CHEMISTRY AND THE ATOMIC THEORY 33

from one compound to another. If a man with his two hands is taken to represent a bivalent atom, it is obviously possible for him to put one hand in his pocket and thus to represent a monovalent atom ; finally, *bringing into play a valency of a different kind*, he might seize an object with his teeth and thus represent a trivalent atom, irrespective of the fact that in more ordinary circumstances the possibility of his so doing might be neglected.

Similarly every atom usually retains the same number of valencies in the various compounds into which it enters. We have never had any reason to suppose that hydrogen is polyvalent; chlorine, bromine, and iodine, which can replace hydrogen atom for atom, are univalent also. Oxygen is usually bivalent, as in water HOH, nitrogen being trivalent, as in ammonia NH_3, or pentavalent, as in ammonium chloride NH_4Cl, and carbon quadrivalent, as in methane CH_4. But the indisputable existence of molecules of nitric oxide NO serves to remind us that oxygen and nitrogen are not always bivalent and trivalent respectively ; again, carbon and oxygen cannot both retain their usual valencies in carbon monoxide CO. Obviously, if such anomalies were of frequent occurrence, the notion of valency, though well founded, would lose much of its usefulness.

23.—CONSTITUTIONAL FORMULÆ.—When the conditions under which a compound is formed are known, it is often possible, by assuming that a minimum of internal dislocation occurs, to make a complete determination of the manner in which the atoms are united in the molecule of the compound and of the number of valencies by which they are held together. This is what is known as establishing the constitution of the compound. The result is open to doubt so long as the constitution is fixed by a single series of reactions. But the doubt is considerably lessened if several series of different reactions point to the same constitution. Representing each saturated valency by a line, we can then represent the compound by a *constitutional formula*, which will possess a wide power of representation with respect to the possible reactions of the compound. For example, we are led by various paths to the opinion that

A.   D

the bonds in the acetic acid molecule are expressed in the formula :

$$\begin{array}{cc} H & O \\ | & \| \\ H\text{—}C\text{—}C\text{—}O\text{—}H \\ | \\ H \end{array}$$

which suggests at once the different *rôles* played by the hydrogen atoms (three being replaceable by chlorine and the fourth by a metal), by the oxygen atoms (the group OH being removed in the formation of acetyl chloride CH_3COCl), and by the carbon atoms themselves (the action of a base KOH on an acetate CH_3COOK splits the molecule up into methane and carbonate).

Constitutional formulæ have taken a position of capital importance in the chemistry of carbon. I shall draw attention to the readiness with which they explain the difference in properties between *isomeric* substances (molecules made up of the same atoms united in different ways [1]) and enable us to predict the number of possible isomers. But I cannot dwell at greater length on the services they have rendered to chemistry, and must content myself with the observation that the 200,000 constitutional formulæ [2] with which organic chemistry is concerned provide just so many arguments in support of the atomic notation and the theory of valency.[3]

24.—STEREOCHEMISTRY.—Once the constitution of the

---

[1] For example, ethanolal HO—$\overset{\overset{\displaystyle H}{\displaystyle |}}{\underset{\underset{\displaystyle H}{\displaystyle |}}{C}}$—$\overset{\displaystyle O}{\overset{\displaystyle \|}{C}}$—H, which possesses both an aldehydic and an alcoholic function, is an isomer of acetic acid.

[2] See Beilstein's Dictionary.

[3] It is moreover possible, and even probable, that, independently of the valencies proper, bonds of a different nature and not so powerful, though equally limited in saturation capacity, may exist between atoms or molecules, giving rise to " molecular compounds " such as double salts or *complex salts*, which are met with more especially in the solid state. Merely to illustrate the possibility of different kinds of union, we may imagine that the ordinary valencies are due to electrostatic attraction and that, in addition, 2 molecules (or even 2 atoms) may attract one another like magnets, which can from astatic systems having no external magnetic action (*cf.* polymerisation by doubling of the molecule, which is frequently observed).

molecule is known, with regard to the modes of union of its component atoms, we may ask ourselves, reasoning as though the molecule were an almost rigid edifice of definite shape what may be the configuration in space of its various atoms. We require to construct in some way a model in three dimensions that will indicate the respective positions of the atoms in space. This new problem, which at first sight would seem to have no meaning (for the valencies might be expected to behave like flexible bonds fixed to a mobile point on the atom, and therefore permitting no definite configuration), has advanced a step towards solution as a result of the splendid work of Pasteur, Le Bel, and van't Hoff, to which I wish to make some reference.

Let us replace successively the four hydrogen atoms in a methane molecule CH_4 by four monovalent groups R_1, R_2, R_3, and R_4, which all differ from each other. If these four groups could occupy any position whatever about the carbon, one single substitution product only could then be obtained. Now two are found, actually very analogous and identical even in certain particulars (they have the same melting points, the same solubility, the same vapour pressure, etc.), but differing sharply in other respects. Their crystals, for instance, which at first sight seem identical, differ in the same way that right-hand gloves differ from left-hand ones, the two kinds being, of course, not mutually replaceable.

Such isomerism is comprehensible if we suppose that the four carbon valencies are attached to the four corners of a practically indeformable tetrahedron. Now there are two non-superimposable ways in which four different objects can be distributed at the corners of such a tetrahedron, and the two arrangements are symmetrical with respect to a mirror as a right-hand glove is to a left-hand one. If, moreover, the tetrahedron is not regular, more than two arrangements producing different solids would be possible (and it should therefore be possible to obtain several di-substitution derivatives having the same formula $CH_2R'R''$, which is contrary to experience).

It therefore seems probable that the molecular edifices are to be regarded, at least approximately, as solid structures, the configuration of which stereochemistry (from δτερεος = solid) aims at determining. Rigidity of the bonds between atoms will appear even more probable when the specific heats of gases (para. 42) have been discussed.

SOLUTION.

25.—RAOULT'S LAWS. — The physical and chemical methods that have been described above are not always sufficient to fix the constitution or even the molecular formula of certain substances. Fortunately a valuable auxiliary is found in the experimental study of dilute solutions.

The formulæ of certain non-volatile substances are difficult to determine. This is the case with numerous "carbohydrates" which by analysis can only be proved to have the formula $C_nH_{2n}O_n$, their chemical properties not always being sufficient to determine n.

Now for a long time it has been definitely known that when a non-volatile substance is soluble in a liquid, for example in water, the solidifying temperature is lower, the vapour pressure less, and the boiling point higher, than is the case with the pure solvent. Thus sea water solidifies at — 2° C. and boils (under normal conditions) at 100·6° C.

But from the restricted study of aqueous solutions of salts it has not been found possible to give precision to these qualitative rules. From his experiments on solutions that, in contradistinction to saline solutions, are not noticeably conductors of electricity and hence are not "electrolytes," Raoult established the following laws (1884) :—

(1) *The influence of each dissolved substance is proportional to its concentration.*[1] The lowering of the freezing point is 5 times greater for a sugar solution that contains 100 grammes of sugar per litre than for one containing only 20 grammes.

(2) *Any two substances exert the same influence when their*

---

[1] This law was stated before Raoult by Wüllner and Blagden, but with reference to electrolytes, the very substances for which it is inaccurate.

CHEMISTRY AND THE ATOMIC THEORY 37

molecular concentrations are equal. More strictly, two solutions (in the same solvent) that in equal volumes contain the same number of gramme molecules have the same freezing point, the same vapour pressure, and the same boiling point.

For the present it is sufficient to recognise the facts embodied in these rules; it may, however, be added that Raoult's more complete statements express the influence due to a given molecular concentration. If n gramme molecules (of any kind) are dissolved in $\mathfrak{N}$ gramme molecules of a solvent, which exerts a vapour pressure p, a solution being thus obtained of vapour pressure p', the relative lowering of vapour pressure, that is $\dfrac{p-p'}{p}$, is sensibly equal to $\dfrac{n}{\mathfrak{N}}$; by dissolving 1 gramme molecule of any substance in 100 grammes of solvent the vapour pressure is lowered by one-hundredth of its value.[1]

With regard to all these laws, it is, of course, understood that the solution must be *dilute*; that is to say, the molecular concentrations must be comparable with those at which gases obey Boyle's Law (more or less of the order of 1 gramme molecule per litre).

It may reasonably be supposed that the above laws are applicable to bodies having molecular formulæ which we do not yet know, as well as to those of known formulæ. If, therefore, a mass m of a substance of unknown formula produces in the boiling point of, for instance, an alcohol solution a variation 3 times smaller than that produced by any of the known gramme molecules when dissolved in the same volume, then the unknown gramme molecule is equal to $3m$. In this way our ability to determine molecular coefficients is enormously increased.

26 —ANALOGY BETWEEN GASES AND DILUTE SOLUTIONS: OSMOTIC PRESSURE.—Raoult's laws, though clear and precise, were nevertheless merely empirical rules. Van't Hoff gave a deeper significance to them when he connected

---

[1] The exact expressions relating to the variation in boiling point and freezing point follow thermodynamically from the expression giving the variation in vapour pressure; further reference is not necessary here.

them with the laws characteristic of the gaseous state, which he was able to show apply also to dilute solutions.

The idea of certain laws being common to all attenuated forms of matter, whether gaseous or in solution, was suggested to him by various botanical researches on osmosis. All living cells are enclosed by a membrane that allows water to pass through but stops the diffusion of certain dissolved substances, the cell gaining or losing water according to the concentration in the aqueous medium in which it is placed (de Vries), which causes the pressure in the interior of the cell to increase or diminsh (it is well known that flowers revive when their stems are placed in pure water but " fade " if the water contains salt or sugar).

Pfeffer succeeded in making indeformable artificial cells which were enclosed by a copper ferrocyanide membrane and which showed the properties described above.[1] When one of these cells, fitted with a manometer and filled with sugar solution, is placed in pure water, the internal pressure steadily rises owing to the entry of water. On the other hand, it is easy to show that no sugar leaves the cell. The ferrocyanide membrane is said to be *semi-permeable*. The excess of internal pressure over the external, moreover, tends to a limit, *proportional to the concentration* for each temperature ; this limit rises when the temperature is raised and returns to its former value (the cell losing water) when the original temperature is reached again. This limiting difference, which is reached when *equilibrium* is attained, is the *osmotic pressure* of the solution.[2]

If then at the bottom of a cylinder we have a sugar solution, above which is pure water, separated from it by a semi-permeable piston, we can concentrate or dilute the sugar solution, according as we press on the piston with a force greater or less than the force just required to balance

[1] Battery jars, of porous porcelain, impregnated with a precipitated membrane of copper ferrocyanide. The cell, previously soaked in water, is filled with a solution of copper sulphate and placed in a solution of potassium ferrocyanide. The precipitated membrane is formed in the pores of the porcelain, from which it cannot escape. The cell is washed, filled with a sugar solution, and sealed with a firm cement.

[2] The order of magnitude is : 4 atmospheres at ordinary temperatures for a 6 per cent. sugar solution.

CHEMISTRY AND THE ATOMIC THEORY 39

the osmotic pressure. Moreover, since this pressure, being proportional to the concentration, is inversely proportional to the volume occupied by the sugar, it would not be apparent, considering only the work required for compression, whether it was being applied to a gas or a dissolved substance.

Van't Hoff, who regarded Pfeffer's experiments from this point of view, was led to the conclusion (van't Hoff's law) that :—

All dissolved substances exert, on a partition that stops them but which allows the solvent to pass, an osmotic pressure equal to the pressure that would be developed in the same volume by a gaseous substance containing the same number of gramme molecules.

Assuming Avogadro's hypothesis, this is the same as :—

Either as a gas or in solution, the same numbers of any kind of molecules whatever, enclosed in the same volume at the same temperature, exert the same pressure on the walls that confine them.

Van't Hoff's theorem, when applied to sugar (which has a gramme molecule of 342 grammes), gives to within 1 per cent. the osmotic pressures measured by Pfeffer. This agreement, though striking, might be accidental. But van't Hoff removed all doubts by showing that his theorem follows necessarily from certain known laws. *Thus, if Raoult's laws are exact, van't Hoff's law must necessarily be so also (and vice versâ).*[1]

27.—IONS--ARRHENIUS'S HYPOTHESIS.—As yet we do not know why it is that a conducting solution, such, for instance, as salt water, does not obey Raoult's laws (and consequently van't Hoff's).

Let us first make clear the nature of this discrepancy ; a mass of salt water containing 1 gramme atom of sodium ($Na = 23$) and 1 gramme atom of chlorine ($Cl = 35\cdot5$), which is 1 gramme molecule 58·5 grammes of sodium chloride,

---

[1] This readily follows from the proof given below (Arrhenius). Let there be, in a region where the gravitational intensity is g and in a vessel free from air, a vertical column of solution in communication with pure solvent through a semi-permeable plug. Let the solution contain n gramme molecules of the dissolved substance (non-volatile) in $\mathfrak{N}$ gramme molecules of solvent. Let equilibrium be reached when the difference in level between the two surfaces is h ; d is the

freezes at a lower temperature than the same volume of solution containing 1 gramme molecule of a non-conducting substance, such as sugar. As the dilution increases, the ratio between the lowerings of freezing point produced by 1 gramme molecule of salt and 1 gramme molecule of sugar increases and tends towards 2, so that, in very dilute solutions, 1 gramme molecule of salt exerts exactly the same influence as 2 gramme molecules of sugar.

This is just what might be expected to happen if, in solution, the salt were partially dissociated into two components that separately obey Raoult's laws, and if, when the dilution is very great, the dissociation were to become complete. We must therefore conclude that the molecules NaCl split up into atoms Na of sodium and Cl of chlorine, and that a very dilute salt solution does not really contain salt, but sodium and chlorine in the form of free atoms. This is the hypothesis that was put forward with such boldness and supported with such brilliance by *Arrhenius* in 1887, he being then a young man of twenty-five.

His conception appeared irrational to many chemists, (mean) density of the vapour, D the very much greater density of the solvent (it is very nearly equal to the density of the solution). Let p' and p be the vapour pressures at the surfaces of the solution and solvent respectively. Then, from the definition of the osmotic pressure P, the pressure at the bottom of the solution is $(p + \mathrm{P})$. The fundamental theorem of hydrostatics, applied to the solution and its vapour, then gives :—

$$p - p' = ghd$$

and $\quad p + \mathrm{P} = p' + gh\mathrm{D}$

whence, eliminating gh, we get approximately

$$\mathrm{P} = (p - p')\frac{\mathrm{D}}{d} = \frac{p - p'}{p} \cdot \frac{p}{d} \cdot \mathrm{D},$$

that is to say, in accordance with Raoult's law stated above,

$$\mathrm{P} = \frac{n}{\mathfrak{N}} \cdot \frac{p}{d} \cdot \mathrm{D}.$$

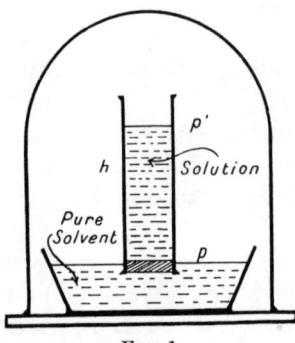

Fig. 1.

Let v be the volume, in the gaseous state at pressure p, of 1 gramme molecule M of the solvent $\left(\text{so that } \frac{p}{d} \text{ is equal to } \frac{p \cdot v}{\mathrm{M}}\right)$; knowing also that $\frac{\mathfrak{N}\mathrm{M}}{n\mathrm{D}}$ is the volume V that is occupied by a gramme molecule of the dissolved substance when in solution, we then have :

$$\mathrm{PV} = pv,$$

which is van't Hoff's law.

and this is all the more curious because, as Ostwald at once pointed out, it was really quite in accordance with well-known facts and also with the binary nomenclature used to represent salts. Thus all the chlorides in solution have certain reactions in common, whatever the metal associated with the chlorine may be, which is readily explained if the same kind of molecule (which can only be the Cl atom) is to be found in all such solutions; with the chlorates, which have a different set of reactions in common, the common molecule would not be Cl but the group ClO_3, and so on.

Disregarding this argument, the opponents of Arrhenius held it to be absurd to assume the existence of free atoms of sodium in water. " It is well known," they said, " that when sodium is placed in contact with water the latter is immediately decomposed with liberation of hydrogen. And further, if chlorine and sodium do co-exist in a solution of salt, simply mixed together like two gases occupying the same vessel, should not means analogous to those applicable to the gaseous state be also available for separating the two elements from each other; by superimposing, for example, above the solution a layer of pure water into which the constituents Na and Cl would certainly diffuse at unequal rates? But attempts to separate them by such means fail, not only in the case of ordinary salt (in which case the rates of diffusion, as an exceptional case, might happen to be equal), but for all electrolytes."

Arrhenius met these objections by insisting upon the fact that the abnormal solutions conduct electricity. This conductivity is explicable if the atoms Na and Cl, which one salt molecule gives on dissociation, are charged with opposite kinds of electricity (in the same way that discs of copper and zinc become charged when separated after previous contact). Speaking more generally, every molecule of an electrolyte can dissociate in the same way into atoms (or groups of atoms), electrically charged, called *ions*. It is assumed that each ion of the same kind, each of the Nations, for instance, in a solution of NaCl, carries exactly the same charge (necessarily equal therefore to the charge of opposite sign carried by the Cl ion, since otherwise the salt solution would not be electrically neutral, as is actually the case). The N atoms which, when neutral, make up 1 gramme atom, constitute, when in the ionic condition, what may be called 1 gramme ion.

When placed in an electric field (such a field is produced when positive and negative electrodes are placed in the salt solution) the positive ions will be attracted towards the negative electrode or *kathode,* and the negative ions will move in a like manner towards the positive electrode or *anode*. A double stream of matter in two opposite directions will thus accompany the passage of electricity. On coming in contact with the electrodes, the ions will lose their charges and acquire other chemical properties at the same instant.

For an ion, which differs by reason of its charge from the corresponding atom (or group of atoms), cannot possess at all the same chemical properties as the latter. As a further result of the charges, diffusion will not be sufficient to effect a separation of the oppositely charged ions. It might well happen, and this is in general the case, that certain ions, say the positive, tend to move faster than the others. They therefore charge positively the region in the liquid where they are in excess ; but at the same time this charge attracts the negative ions, which accelerates their rate of progression and retards at the same time the positive ions. A dog may be more active than a man, but if the dog is held on a leash neither can get along faster than the other.

28.—DEGREE OF DISSOCIATION OF AN ELECTROLYTE.— Finally, the degree of an electrolyte's dissociation into ions can easily be calculated, for each temperature and dilution, if it is assumed, with Arrhenius, that ions obey Raoult's laws as if they were neutral molecules. If a solution containing 1 gramme molecule of salt for each volume V has the same vapour pressure that it would have if it contained $\frac{5}{3}$ gramme molecules of sugar, then it is assumed that it actually contains $\frac{5}{3}$ gramme molecules, necessarily composed of $(1 - \frac{2}{3})$ undissociated gramme molecules of salt and $2 \times \frac{2}{3}$ gramme ions, positive and negative. The *degree of dissociation* $\frac{2}{3}$ is thus found by the application of Raoult's laws.

On the other hand, let us consider a cylindrical column of

CHEMISTRY AND THE ATOMIC THEORY

solution, having a cross-section such that the volume of a section 1 cm. long is the volume we should expect to contain 1 gramme molecule, if we were not aware of its dissociation. As a matter of fact it contains $\frac{2}{3}$ of the ions it would contain if the dissociation were complete. For the same electromotive force the quantity of electricity transmitted per second will therefore be $\frac{2}{3}$ of what would be transmitted at extremely high dilution. More briefly, the conductivity of our cylinder should be, for each centimetre of its length, only $\frac{2}{3}$ of a limiting conductivity that is reached at infinite dilution. Now this is precisely what is found by experiment.

We find the same thing with other salts and at other dilutions; and the degree of dissociation calculated by the application of Raoult's laws is equal to that deduced from the electrical conductivity (Arrhenius's law). Such remarkable concordance, which proves a fundamental connection between properties at first sight as widely different as freezing point and electrical conductivity (a connection so intimate that the one can be predicted when the other is known), clearly lends great support to Arrhenius's theory.

29.—THE FIRST IDEA OF A MINIMUM ELEMENTARY CHARGE.—We have just decided that all the Cl ions in a solution of salt bear the same charge, and we have attributed the difference in chemical properties between the atom and the ion to the existence of this charge. Instead of a solution of sodium chloride, let us now consider one of potassium chloride. Its chemical properties due to the presence of chlorine ions (precipitation with silver nitrate, etc.) are the same as with sodium chloride. The chlorine ions in potassium chloride are therefore probably identical with those of sodium chloride and consequently bear the same charge. Since the solutions are electrically neutral, the sodium and potassium ions must have the same charge also, but with opposite sign. We are thus led step by step to the conclusion that all monovalent atoms or groups of atoms (Cl, Br, I, ClO_3, NO_3 .... and Na, K, NH_4 .... etc.), when they become free in the form of ions, bear the same elementary charge e, positive or negative.

Chlorine ions also possess the same properties, and therefore the same charge, in a solution of barium chloride $BaCl_2$. But in this case a single Ba ion only is formed along with two Cl ions; the charge carried by the Ba ion, which is derived from a bivalent atom, is therefore equivalent, and opposite in sign, to twice the charge on the Cl ion. Similarly the Cu ion, derived from copper chloride $CuCl_2$, carries two elementary charges; so also does the sulphate ion SO_4, but the charges bear the same sign as the Cl ion. The trivalent lanthanum atom in the same way is found to carry three elementary charges when separated from the three chlorine atoms of the chloride $LaCl_3$; and so on.

An important relationship is thus brought out between valency and ionic charge; each valency bond ruptured in an electrolyte corresponds with the appearance of a charge, which is always the same, on the atoms held together by that bond. Moreover, the total charge on an ion must always be an exact multiple of this constant elementary charge, which is, indeed, an actual *atom of electricity*.

The above view is completely in accordance with the knowledge we have gained from the careful study of electrolysis. I feel that some account of this is very desirable, since, in my opinion, the usual methods of presenting it are not at all satisfactory.

30.—THE CHARGE CARRIED BY A GRAMME ION—ELECTRICAL VALENCY.—When two electrodes are placed in an electrolyte, changes are at once observed to take place in their immediate neighbourhood. Bubbles of gas, solid particles, or drops of liquid make their appearance on the electrode surfaces, rising or sinking according to their density and so tending to contaminate regions which the passage of the current alone would perhaps have left unaltered.

Complications of this kind can be avoided by making the current follow a curvilinear course; for example, in the way indicated in the diagrammatic sketch shown below. The electrolyte is divided into two parts, contained in two beakers, an electrode being fixed in each. The two beakers are connected by means of a siphon containing a column of liquid, through which the current must pass but which cannot be entered by substances rising or falling from the electrodes. Precautions are taken, moreover, to prevent the loss of these substances.

It is then easy to show conclusively that the mere passage of the current does not affect the electrolyte; we have only to remove the siphon tube after the passage of a certain quantity Q of electricity (which can easily be measured by a galvanometer) and to analyse the liquid contained in it, when the solution will be found to have undergone no change.

At the same time the substances produced from the rest of the materials concerned in the experiment have been separated into two compartments. It will be possible to

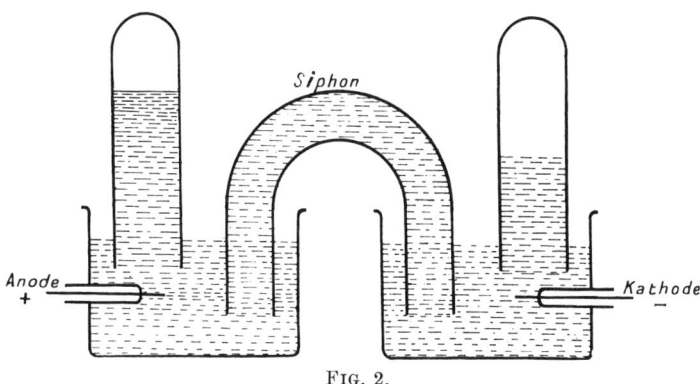

Fig. 2.

analyse the contents of each compartment (including the products formed on the electrodes) and to determine the number of gramme atoms of each kind found therein.

Let us suppose that a solution of salt has been electrolysed. In the kathode compartment we shall find, first, part of the kathodic material (supposing the kathode to have been attacked); then (in terms of hydrogen, oxygen, chlorine, and sodium) the materials that constitute a salt solution; and, finally, an excess of sodium, so that the total composition of the compartment can be expressed by a formula such as :—

$$(\text{kathode}) + a(2\text{H} + \text{O}) + b(\text{Na} + \text{Cl}) + x\text{Na}.$$

It must be clearly understood that we are here dealing with

a formula expressing the gross composition of the contents of the compartment, independently of any hypothesis as to the particular compounds that may be present. It is immaterial that the $2\,a$ gramme atoms of estimated hydrogen are partly in the form of gaseous hydrogen and partly combined in water or sodium hydroxide molecules; we are concerned with their total number only.

At the same time, since no matter has been lost, the *total* formula for the anode compartment must be :—

(original anode) $+\, a'(2\text{H} + \text{O}) + b'(\text{Na} + \text{Cl}) + x\text{Cl}$,

the number of gramme atoms x of chlorine present in excess in this compartment being equal to the number of gramme atoms of sodium present in excess in the kathode compartment.

Thus, by causing the quantity of electricity Q to pass through the solution, x gramme molecules of salt have been *decomposed* into sodium and chlorine, which have been obtained *separate*, one component in each of the two compartments.

No matter how the experimental conditions are varied (dilution, temperature, nature of the electrodes, current density, etc.) the passage of the same quantity of electricity always decomposes the same number of gramme molecules. Thus, when twice, three times or four times more electricity is passed through, twice, three times or four times more electrolyte is decomposed. The quantity of electricity F (equal to 96,500 coulombs) which is accompanied in its passage through the solution by the decomposition of 1 gramme molecule of salt, is often called a *faraday*, after Faraday, who first observed this *exact* proportionality.

It is easy to see that the charge carried by a gramme ion must be 1 faraday exactly. If not, let F′ be its charge, supposed to be different from F. Let 1 faraday be caused to pass through the electrolyte; if m gramme atoms of sodium pass across a section midway between the electrodes, in the direction of the kathode (carrying with them mF′ positive faradays), $(1-m)$ gramme atoms of chlorine must cross the section in the opposite direction (carrying with

CHEMISTRY AND THE ATOMIC THEORY

them $(1 - m)F'$ negative faradays). The faraday passed through is therefore equal to $(m + 1 - m)F'$; that is, to the charge F' carried by 1 gramme ion of sodium or of chlorine.

If, instead of sodium chloride, we electrolyse a solution of potassium chloride KCl, we find, by exactly similar experiments, that the passage of 1 faraday again decomposes 1 gramme molecule. As we should expect from chemical reasons, the gramme ion of chlorine bears the same charge in potassium chloride as in sodium chloride. And in the same way it may be shown that every monovalent ion carries with it 1 faraday, positive or negative. This is so, for instance, for the hydrogen ion H^+, which is characteristic of acids, and for the hydroxyl ion OH^-, characteristic of bases.

It is also found, as might be expected, that 2 faradays must pass in order to bring about the decomposition of 1 gramme molecule of barium chloride, $BaCl_2$, the ion Ba^{++} thus bearing two elementary charges. And we find that the passage of 2 faradays decomposes 1 gramme molecule of copper sulphate $CuSO_4$ (producing in the anode compartment an excess of 1 gramme atom of the group SO_4), so that the ions SO_4^{--} and Cu^{++} each carry twice the charge borne by the ions Cl^- and Na^+; and so on.

In short, all monovalent ions carry the same elementary charge e, either positive or negative in sign, e being equal to the quotient $\dfrac{F}{N}$ obtained by dividing the faraday by Avogadro's number, in accordance with the equation

$$F = Ne\ ;$$

and all polyvalent ions carry as many of these charges as they have valencies.

It does not appear to be possible to obtain a sub-multiple of this elementary charge, which thus possesses the essential characteristic of an atom, as Helmholtz first pointed out in 1880. *It is indeed an atom of electricity.* Its absolute value will be known when we succeed in obtaining N.

Some indication of the vastness of the charges transported by the ions may be given with advantage. It can be shown, by the application of Coulomb's law, that if it were possible to obtain two spheres each containing 1 milligramme atom of monovalent ions, placed 1 centimetre apart, they would repel or attract each other (according to the signs of the two lots of ions), with a force equal to the weight of 10^8 metric tons. This is sufficient explanation of the fact that a separation of the Na and Cl ions present in a solution of salt to any great extent, such as was demanded of Arrhenius, cannot be effected, either by spontaneous diffusion or in any other way.

An Upper Limit to Molecular Size.

31.—DIVISIBILITY OF MATTER.—Up to the present it has been my endeavour to collect together the arguments that led to a belief in an atomic structure for matter and electricity and yielded us the ratios between the weights of the atoms, supposing that they exist, before any idea as to the absolute values of these magnitudes had been formed.

I need scarcely point out that these magnitudes elude direct observation. As far as the subdivision of matter has been pushed up to the present, there has been no indication that any limit has been approached and that a granular structure lies beyond the limits of direct perception. A few examples will be useful in reminding us of this extreme divisibility.

Gold workers prepare gold-leaf having a thickness of only *one ten-thousandth of a millimetre or, more shortly, one-tenth of a micron*. These leaves, which are familiar to all of us and which are transparent and transmit green light, appear nevertheless to be continuous in structure; we cannot push the subdivision any further, not because the gold ceases to be homogeneous, but because it becomes more and more difficult to manipulate the thin leaves without tearing them. If gold atoms do exist, their diameter is therefore less than one-tenth of a micron ($\cdot 1\mu$ or 10^{-5} cms.), and their mass must be less than the mass of gold that fills a cube of that diameter; that is, it must be less than the hundred-thousandth of a milligramme (10^{-14} grs.). The mass of the hydrogen atom, which is, as we have seen, about 200 times lighter, is thus so minute that certainly more than 20 million atoms

CHEMISTRY AND THE ATOMIC THEORY 49

are needed to make up 1 milligramme ; in other words, its mass is less than $\frac{1}{2} \times 10^{-16}$ grammes.

Microscopical examination of various bodies enables us to go much further, particularly in the case of strongly fluorescent substances. Indeed, I have satisfied myself that a solution of fluorescein containing one part in a thousand, illuminated at right angles to the microscope by a parallel beam of very intense light (the ultra-microscopic arrangement), still shows a uniform green fluorescence in volumes of the order of a cubic micron. The mass of the bulky fluorescein molecule, which we know (from its chemical properties and Raoult's laws) to be 350 times heavier than the hydrogen atom, is therefore certainly less than the one-thousandth part of the mass of a cubic micron of water. This means that the hydrogen atom certainly weighs less than the one-thousandth part of one-thousandth part of a milligramme. Briefly, the hydrogen atom has a mass less than 10^{-21}. Avogradro's number N is therefore greater than 10^{+21} (so that there are more than 1,000 milliards of milliards of molecules in a gramme molecule).

Since the hydrogen atom weighs less than 10^{-21} grammes, the water molecule, which is 18 times as heavy, must weigh less than 2×10^{-20} grammes. Its volume is therefore less than 2×10^{-20} cubic centimetres (since 1 cubic centimetre of water contains 1 gramme) and its *diameter* is less than the cube root of 2×10^{-20} ; less, that is to say, than four hundred thousandths of a millimetre ($\frac{1}{4} \times 10^{-6}$ cms.).

32.—THIN FILMS.—The study of "thin films" leads us still further. During the blowing of a soap bubble we often notice, in addition to the familiar brilliant colours, small round black spots with well-defined edges. These spots might be taken for holes, and their appearance is almost immediately followed by the rupture of the bubble. They may be readily observed while washing the hands, on a film of soapy water stretched over the space between the thumb and forefinger. When this film is held vertically the water in it gradually drains to the bottom, while the upper part becomes thinner and thinner, which process can be followed by the colour changes that occur. After it has become purple and then pale yellow, the appearance of the black spots will soon be noticed; they run together, forming a black space, which may fill a quarter of the height of the film before it breaks. If this rough experiment is repeated with certain precautions, the thin films being produced on a fine framework inside a case to protect them from evaporation, it is possible to maintain these black surfaces in equilibrium for several months (Sir J. Dewar) and so to observe them at leisure.

In the first place, these black spots are not holes, for it is easy to show, as was done by Newton, who first studied them, that, though black by contrast, they nevertheless reflect light, and also that new round, sharp-edged spots ultimately appear within the original spots; the new spots are still darker and hence thinner, and *also* reflect feeble images of bright objects, such as the sun.

It is possible to measure [1] the thickness of the black spots, and it has been found that the blackest and thinnest have a thickness of about $4 \cdot 5 \times 10^{-7}$ cms. The others have approximately double this thickness, which is somewhat remarkable.[2]

The films produced by the spreading of oil drops on a water surface may become even thinner than the black spots on soap bubbles, as Lord Rayleigh has shown. It is known (and the fact can easily be verified) that small pieces of camphor thrown upon quite pure water commence to dart about in all directions on the surface of the water (for the solution of the camphor is accompanied by a considerable lowering of the surface tension, with the result that each piece is continually being urged into regions where solution is less active). This phenomenon is not observed if the water surface is greasy (and has in consequence a surface tension much lower than that of pure water). Lord Rayleigh has attempted to determine the

[1] The simplest method is to determine their reflecting power and to apply the classic theory of thin films (*Ann. de Phys.*, J. Perrin, 1918, and Wells, 1921).
[2] I have shown that with certain precautions it is possible to produce films made up of more than a hundred layers, with sharp edges and of uniform thicknesses, which are whole number multiples of the thinnest, black film (*Ann. de Phys.*, J. Perrin, 1918, and Wells, 1921).

CHEMISTRY AND THE ATOMIC THEORY 51

weight of the smallest drop of oil that, when placed on the surface of a large basin of very pure water, is found to be just sufficient to prevent the movement of the camphor at all points on the surface. This weight was so small that the thickness of oil thus spread over the surface of the water could not have reached two-thousandths of a micron.

Devaux has made a comprehensive study of these thin films of oil, which he very happily compares with the black spots on soap bubbles. Thus, when a drop of oil spreads upon water, an iridescent film is seen to form, in which black circular, sharp-edged spots soon appear. Within them the liquid surface is still covered with oil, since it still possesses the properties described by Lord Rayleigh. But this oil has not yet reached its maximum extension; by allowing a drop of a dilute standard solution of oil in benzene (which evaporates rapidly) to fall on a large water surface, Devaux has obtained an oil film free from thick spots and with sharply defined edges. He demonstrated the presence of the oil film, not with camphor (which moves on the film as if it were pure water), but with powdered talc. When sprinkled onto pure water with a sieve, this powder is easily shifted by blowing horizontally on the liquid, and collects on the opposite side of the basin, where the surface is dimmed. But its motion is stopped by the edges of the oil film and marks their limits. In this way it is possible to measure the surface of the film with an accuracy bordering on the one-hundreth of a millimicron. The corresponding thickness is very little more than a millimicron ($1 \cdot 10 \mu\mu$ or $1 \cdot 1 \times 10^{-7}$ cms.).

It must be borne in mind that in these measurements it is assumed that the material of the film is uniform in thickness; and that, after all, it is not certain, considering only the facts at present known, that the films have not a reticular or fine-meshed structure, like a spider's web, which at a distance may appear homogeneous.

It seems more probable, however, that the thin films are nowhere thicker than the mean measured thickness, and that the maximum diameter possible for the oil molecule is in consequence of the order of a millimicron. It will be considerably less for the constituent atoms ; the maximum mass possible for a molecule of oil (glycerine tri-oleate, $C_{57}H_{104}O_6$) would be of the order of one thousand-millionth of one thousand-millionth of a milligramme, and the mass of a hydrogen atom, which is nearly a thousand times less, would be of the order of a millionth of a trillionth of a gramme (10^{-24} grs.).

We may summarise this discussion by stating that the different atoms are certainly less than a hundred-thousandth (perhaps a millionth) of a millimetre in diameter, and that the masses even of the heaviest (such as the gold atom) are certainly less than a thousandth of a trillionth of a gramme.

However small these superior limits, which mark the actual boundaries of our direct perception, may appear, they may nevertheless be vastly greater than the actual values. Certainly when we review, as has been done above, all that chemistry owes to the conceptions of atom and molecule, it is hard to doubt at all seriously the existence of such elements in matter. But at present we are not in a position to decide whether they lie just on the threshold of the directly perceptible magnitudes or whether they are so inconceivably small that we must regard them as infinitely removed from our sphere of cognisance.

This is a problem which, once stated, should prove a powerful incentive to research. The same ardent and disinterested curiosity that has led us to weigh the stars and map out their courses urges us towards the infinitely small as strongly as towards the infinitely large. Striking advances already made give us the right to hope that our knowledge of both atoms and of stars may become equally complete.

CHAPTER II

MOLECULAR AGITATION

THE transference of matter that occurs during solution or diffusion has led us to suppose that the molecules in a fluid are in incessant motion. By developing this idea in conformity with the laws of mechanics, *which are assumed to be applicable to molecules,* an important collection of propositions has been brought together under the name of the *kinetic theory.* This theory has shown great fertility in the explanation and prediction of phenomena, and was the first to yield a definite indication of the absolute values of the molecular magnitudes.

MOLECULAR SPEEDS.

33.—MOLECULAR AGITATION A PERMANENT CONDITION.— As long as the properties of a fluid appear to us invariable, we must suppose that molecular agitation in that fluid neither increases nor decreases.

Let us endeavour to define this rather vague proposition. In the first place (as is shown by experiment), equal volumes contain equal masses, that is to say, equal numbers of molecules. More accurately, if n_o denotes the number of molecules that should be found in a certain volume if their distribution were absolutely uniform, then, if n is the number actually found at a given moment, the *fluctuation* $n - n_o$, which varies from instant to instant with the random motion of the molecular agitation, will be of less importance the greater the volume considered. In practice, it is quite negligible for the smallest volumes observable.

Similarly there is practical equality, in any arbitrary portion of the fluid, between the number of molecules moving with a certain velocity in one direction and the number moving with the same velocity in the opposite direction. More generally, if we consider a large number of molecules, *taken at random* at a given moment, then the projection of all the molecular speeds onto any arbitrary axis (in other words, their resultant along that axis) will have a mean value of zero ; no particular direction will be privileged.

Similarly, the aggregate energy of motion or kinetic energy associated with a given portion of matter will experience none but the most insignificant fluctuations for those portions that can be observed. More generally, if we consider at any given moment two groups of equal (and sufficiently large) numbers of molecules, which have been separately chosen at random, then the sum of the kinetic energies of the molecules is practically the same for the two groups. This comes to the same thing as saying that the molecular energy has a fixed mean value W, which is always found to be the same on taking the mean, at any given moment, of the molecular energies of molecules chosen at random in any number, so long as it is large.

The same value W would be obtained if we took the mean of the energies possessed by the same molecule at different instants (a large number must be considered) distributed at random over a considerable period of time.[1]

The above remarks hold for every kind of energy that can be attributed to the molecule. They apply particularly to the kinetic energy of translation $\frac{1}{2}mV^2$, m being the mass and V the velocity of the centre of gravity of the molecule. The mass being constant, if there is a definite mean value w for this energy of translation, there will be a definite mean value U^2 for the square of the molecular velocity.

Similar remarks apply to all definable properties of the molecules in a fluid. There is, for example, a definite value G for the mean molecular velocity. This value is not U, as will be obvious when it is recalled that the mean $\frac{a+b}{2}$ of

---

[1] Clearly this mean value W' is the same for any two molecules (which do not differ in their capacity for acquiring energy) ; thus, let the energies of a large number of molecules p be estimated at q successive instants (q being very great). The sum of the energies thus measured may be written either q times pW or p times qW', which shows that W is equal to W'.

MOLECULAR AGITATION

two different numbers a and b is always less than the square root of the mean of the squares of these numbers, $\frac{a^2 + b^2}{2}$.

U is sometimes called the mean quadratic velocity.

Maxwell showed that when the mean square U^2 is known, the mean speed G follows from the probability law that fixes the proportion of molecules that have a certain velocity at each instant.

He arrived at these results, which are of great importance in our study of the permanent condition of molecular agitation, by assuming that the proportion of the molecules having a definite velocity component in a given direction is the same both for all the molecules together and for that proportion of their number known independently to possess another definite component in a perpendicular direction. (More briefly, if we consider two walls at right angles, and if we suppose that at a given moment a molecule is moving with a velocity of 100 metres per second towards the first wall, then, according to Maxwell, we can gain *no* information from this fact as to the probable value of its velocity towards the second wall.) This hypothesis as to the distribution of velocities, which is probable though by no means certain, is justified by its results.

By a calculation involving no other hypothesis, and which may therefore be omitted in detail without affecting our attitude towards the phenomena under discussion, it is possible to determine completely the *velocity distribution*, which is the same for all fluids in which the mean square of the molecular velocity has the same value U^2. In this way it is possible to calculate the mean velocity G, which is found to be less than U and to be approximately equal to $\frac{12}{13} U$.[1]

---

[1] To be precise, out of $\mathfrak{N}$ molecules, the number dn of them that have a component along Ox lying between x and $x + dx$ is given by the equation :—

$$dn = \mathfrak{N}\sqrt{\frac{3}{2\pi}} \cdot \frac{1}{U} \cdot e^{-\frac{3}{2} \frac{x^2}{U^2}} \cdot dx,$$

and, moreover, we have :—

$$G = U\sqrt{\frac{8}{3\pi}}$$

56 ATOMS

34.—CALCULATION OF THE MOLECULAR VELOCITIES.—If the fluid is gaseous, a simple theory gives, with considerable accuracy, the value of the mean square U^2 of the molecular velocity, from which the mean velocity and velocity distribution follow.

We have already decided that the pressure exerted by a gas is due to the continual impact of the molecules against the walls of the containing vessel. In developing this idea we will assume that the molecules are perfectly elastic. Then, in order to find their velocity, it is merely necessary to calculate the constant pressure supported by unit surface of a rigid wall uniformly bombarded by a regular stream of projectiles, which move with equal and parallel velocities and which rebound from the wall without gain or loss of energy. This is a mechanical problem, into which no physical difficulties enter; I shall therefore omit the calculation (which is, moreover, simple) and give only its solution, namely, that the pressure is equal to twice the product of the velocity component perpendicular to the wall (which component changes its sign during the impact) into the total mass of the projectiles striking unit surface in unit time.

Under equilibrium conditions, the assemblage of molecules near a partition may be regarded as a large number of streams of this kind, moving in all directions and without the least influence on each other if the molecules occupy but little of the space they move in (this is the case when the fluid is gaseous). Let x be the velocity perpendicular to the partition for one of these streams and q the number of molecular projectiles per cubic centimetre; then qx projectiles per second, of total mass qxm, will strike each square centimetre of the partition, which will in consequence be acted on by a partial pressure $2qmx^2$. The sum of the pressures due to all the streams will be $2\dfrac{n}{2}mX^2$, where X^2 is the mean square of the component x, and n is the total number of molecules per cubic centimetre (of which only a fraction are moving towards the partition). Hence, since the mass $m \times n$ of each unit volume is the density (absolute) of the gas, we see that the pressure p is equal to the product $X^2 d$ of the density by the mean square of the velocity parallel to an arbitrary direction. Incidentally, we find at the same time that the mass of gas which strikes a square centimetre of the partition per second is equal to $X'd$, where X' is the mean value of those of the components x that are directed towards the partition; since X' (which becomes doubled or tripled when the velocities are doubled or tripled) is proportional to the mean speed G, this mass is proportional to Gd (a result we shall use later on).

The square of a velocity, that is to say, the square on the diagonal of a parallelepiped constructed from three rectangular components, is equal to the sum of the squares on the three components, and hence the mean square U^2 is equal to $3 X^2$ (the three rectangular projections having by symmetry the same mean square). The pressure p, equal to $X^2 d$, is therefore also equal to $\frac{1}{3} U^2 d$ or $\frac{1}{3} \cdot \frac{M}{v} \cdot U^2$, where M is the mass of gas occupying volume v.

We have thus established the equation

$$3pv = MU^2,$$

which may be written

$$\frac{3}{2} pv = \frac{MU^2}{2},$$

and may be stated as follows :—

For any given mass of gas, the product of the volume by the pressure is equal to two-thirds of the energy of translation associated with the molecules in the mass.

We know, moreover (Boyle's Law), that at constant temperature the product pv is constant. The molecular kinetic energy is therefore, at constant temperature, independent of the rarefaction of the gas.

It is now easy to calculate this energy, as well as the molecular velocities, for any gas, at any temperature. The mass M may be taken equal to the gramme-molecule. Since all gramme molecules occupy the same volume under the same pressure (para. 18), which means that the product pv is the same for all, we see that, in the gaseous condition :—

The sum of the energies of translation of the molecules contained in a gramme-molecule is the same for all gases at the same temperature.

At the temperature of melting ice this total energy is 34,000,000,000 ergs.[1] Expressed in other terms, the work done by the stoppage, at this temperature, of all the molecules contained in 32 grammes of oxygen or 2 grammes of hydrogen would be sufficient to raise 350 kilogrammes through 1 metre ; this shows what a reserve of energy lies in molecular motion.

Knowing the energy $\frac{MU^2}{2}$ of a known mass M, we can at once obtain U and in consequence the mean velocity G. Again, at the temperature of melting ice, the kinetic energy for oxygen (M = 32) is the same as if, supposing that all the molecules were stopped, the mass considered had as a whole the velocity U of 460 metres per second. The mean velocity G, which is slightly less, is 425 metres per second. This is not much less than the speed of a rifle bullet. In hydrogen (M = 2) the mean velocity rises to 1,700 metres ; it falls to 170 metres for mercury (M = 200).

35.—ABSOLUTE TEMPERATURE (PROPORTIONAL TO THE MOLECULAR ENERGY).—The product pv of the volume by the pressure, which is constant for a given mass of gas at a fixed temperature (Boyle), changes in the same way for all gases as the temperature is raised (Gay-Lussac). In point of fact, it increases by $\frac{100}{273}$ of its value on passing from the temperature of melting ice to that of boiling water. As we know, this enables us to define (by means of the gas thermometer) a *degree* of temperature as being the increment of temperature that raises the product pv (or simply the pressure if we work at constant volume) for any gas by $\frac{1}{273}$ of the value it has at the temperature of melting ice (so that

---

[1] For each gramme molecule occupies 22,400 cubic centimetres when the pressure corresponds to 76 centimetres of mercury, which gives for the product $\frac{3}{2}pv$ the value 34×10^9 in C.G.S. units.

MOLECULAR AGITATION 59

there are 100 such degrees between the temperature of melting ice and that of boiling water).

Now we have seen above that the molecular energy is proportional to the product pv. Thus for a long time we have unwittingly been accustomed to mark equal steps on the temperature scale by equal increments of molecular energy, the increment of energy per degree being $\frac{1}{273}$ of the molecular energy at the temperature of melting ice. As we have already shown (para. 4), heat and molecular agitation are in reality the same thing viewed under different magnifications.

Since the energy due to molecular agitation cannot become negative, the *absolute zero* of temperature, corresponding to molecular immobility, will be reached 273 degrees below the temperature of melting ice. *Absolute temperature*, which is proportional to the molecular energy, is reckoned from this zero; the absolute temperature of boiling water, for example, is 373 degrees absolute.

It appears that for any gaseous material the product pv is proportional to the absolute temperature T; this gives us the equation for a perfect gas:—

$$pv = rT.$$

Let R be the particular value,[1] independent of the nature of the gas, that r takes when the quantity of gas chosen is a gramme molecule. If the quantity considered contains n gramme molecules, the preceding equation can be written

$$pv = nRT.$$

Finally, since the molecular kinetic energy is, as we have seen, equal to $\frac{3}{2}pv$, we can write, for a gramme molecule M:—

$$\frac{MU^2}{2} = \frac{3}{2}RT.$$

36.—PROOF OF AVOGADRO'S HYPOTHESIS.—It appears that any two gramme molecules, considered in the gaseous

[1] From the fact that 1 gramme molecule occupies 22,400 cubic centimetres at atmospheric pressure at the temperature of melting ice (T = 273° A), we get that R is equal to 83 2 × 10³ C.G.S. units

state at the same temperature, each contain the same amount of molecular energy of translation $\left(\frac{3}{2} RT\right)$. Now, according to Avogadro's hypothesis, two such quantities of gas each contain the same number of molecules N. At the same temperature, the molecules of the respective gases therefore possess the same mean energy of translation w (equal to $\frac{3}{2} \cdot \frac{R}{N} \cdot T$). The hydrogen molecule is 16 times lighter than the oxygen molecule, but it moves on the average 4 times more quickly.

In a gaseous mixture each molecule, of whatever kind, has this same mean energy. For we know (from the law of gaseous mixtures) that each of the gaseous masses mixed in a receptacle exerts on its walls the same pressure that it would exert if present alone. From the expression giving the partial pressure of each gas (which we may treat exactly as in the case of a single gas), it follows that the molecular energies must be the same before and after mixing. Whatever the nature of the constituents of a gaseous mixture, any two molecules chosen at random will possess the same mean energy.

This *equipartition of energy* between the various molecules of a gaseous mass, worked out above as a consequence of Avogadro's hypothesis, can be demonstrated without reference to that hypothesis, if it is assumed, as has already been done, that the molecules are perfectly elastic.

The demonstration is due to Boltzmann.[1] He considered a gaseous mixture containing molecules of two kinds, having masses m and m'. If we are given the velocities (and hence the energies) of the two molecules m and m' before an impact and the direction of the line joining their centres after impact, the laws of mechanics enable us to calculate what their velocities will be after impact. The gas is, moreover, in a state of internal equilibrium ; the disturbing effect on the distribution of velocities caused by one kind of impact must therefore be compensated continuously by impacts of the " opposite " kind (the quantity of motion of the colliding

[1] "Theorie cinétique," chap. 1.

molecules being just the same as before, but of opposite sign). Boltzmann then succeeded in showing, *without any further hypothesis*, that this continuous compensation implies equality between the mean energies of the molecules m and m'. Thus the law of gaseous mixture requires that these mean energies should remain the same for the gases when separate (which is the result arrived at above).

Since, moreover, we have shown that the total molecular energy is the same for masses of different gases occupying the same volume under the same conditions of temperature and pressure, it follows that such masses must contain the same number of molecules, which is Avogadro's hypothesis. Justified by its results but introduced nevertheless in a somewhat arbitrary manner (para. 13), the hypothesis now finds its logical basis in Boltzmann's theory.

37.—EFFUSION THROUGH SMALL ORIFICES.—The values derived above for the molecular velocities cannot as yet be verified directly. But the values obtained for gas pressure also give us a quantitative expression of two quite different phenomena, which provides us with a valuable check upon the theory.

One of these phenomena is the *effusion*, or progressive passage, of a gas through a very small opening pierced in a very thin partition enclosing the gas. To understand the mechanism of this effusion we must bear in mind that the mass of gas striking per second against a given element of the partition is proportional to the product of the mean molecular velocity G into the density of the gas. Now suppose that this element of the partition is suddenly removed; the molecules which were about to strike against it will now disappear through the opening. The initial loss will be proportional to Gd; it will remain constant if the opening is so small that the balance between the various molecular motions is not disturbed to any great extent.

The *mass* thus effused being proportional to Gd, its *volume* under the pressure in the enclosure must be proportional to the molecular velocity G or, which is the same thing, to the mean quadratic velocity U (equal to $\frac{13}{12}$ G).

Since, in short, at constant temperature the product MU^2 is independent of the nature of the gas, it follows that :—

The volume effused in a given time must be inversely proportional to the square root of the molecular weight of the gas.

The various common gases have been found to obey [1] this law. Hydrogen, for instance, effuses 4 times more rapidly than oxygen.

38.—WIDTH OF SPECTRAL LINES.—The phenomenon of effusion provides us with a means for checking the *ratios* of the molecular speeds of the various gases but leaves indeterminate the *absolute* values of those speeds, which, according to what has been said above, should reach several hundred metres per second.

Now attention has recently been directed towards a phenomenon that has no apparent connection with the pressure exerted by gases, but which again enables us to calculate the speeds of molecules, supposing that they exist, and which yields exactly the same values.

It is well known that the electric discharge causes rarefied gases to glow. When examined with the spectroscope, the light emitted from " Geissler tubes " in action is seen to be resolved into fine " lines," each corresponding to a single homogeneous beam of light, which may be compared with sound of a definite pitch. Nevertheless, if the contrivance for splitting up the light is made sufficiently powerful (by the use of diffraction gratings and, best of all, of interferometers), the finest lines are found ultimately to have an appreciable thickness.

That this should be so was predicted by Lord Rayleigh, from the following very ingenious considerations. He supposed that the light *emitted* by each vibrating centre (atom or molecule) is in reality homogeneous ; but such centres being always in motion, the light perceived has a longer or shorter period, according as the vibrating centre is approaching or receding.

---

[1] Once established, this law enables us to determine unknown molecular weights ; if it takes, for instance, 2·65 times as long to empty by effusion the same enclosed space to the same extent when it contains radium emanation as when it contains oxygen, the molecular weight of the emanation can be found by multiplying the molecular weight of oxygen, 32, by $(2·65)^2$, or about 7.

MOLECULAR AGITATION

In the case of sound we are familiar with a phenomenon of this kind. It is known that the sound of a motor horn, emitted at a pitch that is obviously fixed, appears to alter when the motor is in motion. It becomes sharper as the motor approaches (for then more vibrations are perceived per second than are emitted in the same time), and suddenly becomes deeper as soon as it has passed (for then fewer vibrations are received). A simple calculation shows that if v is the velocity of the source of sound and V that of sound itself, the pitch of the sound perceived may be obtained by multiplying or dividing the real pitch by $\left(1 + \dfrac{v}{V}\right)$, according as the source is approaching or receding. (This will cause a sudden variation, of the order of a third, when the source passes us.)

The same considerations apply to light, in which connection they are known as the Doppler-Fizeau principle. In the first place this principle explains why, with ordinarily good spectroscopes, the lines characteristic of the metals found in different stars are sometimes all seen to be displaced slightly towards the red (receding stars) and sometimes towards the violet (stars that are approaching us). The velocities of the stars measured in this way are on the average of the order of 50 kilometres per second.

But with better instruments velocities of several hundred metres per second can be detected. If the bright capillary section of a Geissler tube containing mercury vapour and immersed in melting ice is observed at right angles to the electric force,[1] the nature of the light perceived proves the existence of an enormous number of atoms moving in all directions with velocities of the order of 200 metres per second ; rigorously homogeneous light can no longer be perceived, and an apparatus of sufficiently high dispersive power will reveal a diffuse band instead of an indefinitely thin line. The mathematical theory enables us to calculate the mean molecular velocity corresponding to the broadening

---

[1] Neglecting the increase in velocity in its own direction that this force can communicate to the luminous centre, if the latter is charged (Stark has established the Doppler effect in the positive " canal " rays in Crookes' tubes).

observed. It then only remains to be seen whether this velocity agrees with that derived, according to the theory discussed above, from a knowledge of the gramme molecule and the temperature.

Experimental work has been carried out by Nicholson, and also by Fabry and Buisson, whose experiments were more accurate and in some cases more numerous. Their results leave no room for doubt; the velocities calculated by the two methods agree to within nearly 1 per cent. (Qualitatively, a line is broader the smaller the molecular mass of the glowing gas and the higher its temperature.)

Having once established this remarkable agreement for certain gases and for particular lines, it will be legitimate to regard it as still holding in cases where either the molecular mass or the temperature is unknown and to use it for determining the latter magnitudes. In this way Buisson and Fabry showed that in a Geissler tube in action containing hydrogen the luminous centre is the hydrogen atom and not the molecule.[1]

Molecular Rotations and Vibrations.

39.—The Specific Heat of Gases.—Up to the present we have confined our attention to the translatory movements of the molecules. But the molecules probably spin round while they move, and if they are not rigid, other more complicated motions may occur.

Consequently, when the temperature is raised, the energy absorbed during the heating of 1 gramme molecule of a gas must be greater than the increase in molecular energy of translation, which we know to be equal to $\frac{3}{2}RT$. For each rise of 1° C., at constant volume (in which case all the energy acquired by the gas is communicated by heating alone and

---

[1] Following up this brilliant piece of research, Buisson and Fabry have determined the temperature of the nebulæ from the observed broadening of the hydrogen lines : having done that they have been able to determine the atomic weight, 3, of the body (nebulium), which, in the same nebulæ, produces certain lines belonging to no known terrestrial element. In this way the atom of a simple substance has been discovered and weighed in regions so far distant that light from them takes centuries to reach us !

none by work done in compression), the quantity of heat absorbed per gramme molecule of the gas (specific heat at constant volume) will therefore be greater than or equal to $\frac{3}{2}$ R C.G.S. units of energy (ergs); that is, to 2·98,[1] or approximately 3 calories.

This is a very remarkable limitation. A single well-established case where the heat lost by 3 grammes of water in cooling 1 degree raises the temperature of 1 gramme molecule of a gaseous substance by more than 1 degree would be sufficient to jeopardise the kinetic theory. But no such case has ever been recorded.

40.—MONATOMIC GASES.—The question arises whether the molecular specific heat at constant volume (which we shall call c) can actually fall to the above inferior limit of 3 calories. In any case where this occurred, the inference would be that not only does the internal energy of the molecule remain unchanged as the temperature rises but that its rotational energy also remains constantly at zero, so that two molecules striking against each other must behave like two perfectly smooth spheres, there being no frictional effect at the moment of impact.

If any molecules should happen to possess this property, they might be expected to be molecules consisting of single atoms. Such molecules are found in mercury vapour, and consequently the determination of c for that substance is of particular interest. Experiments carried out by Kundt and Warburg gave the value 3 exactly. (The same result has been obtained for the monatomic vapour of zinc.)

Furthermore, Rayleigh and Ramsay have discovered certain gaseous, chemically inactive simple substances (helium, neon, argon, krypton, xenon), which, owing to their inactivity, had remained hitherto unknown to chemists. These bodies, which cannot be made to combine with any other substances, are probably composed of atoms of zero valency which are no more able to combine among them-

---

[1] For $\frac{3}{2}$R is equivalent to $12 \cdot 5 \times 10^7$ ergs, or (since a calorie is equivalent to $4 \cdot 18 \times 10^7$ ergs) 2·98 calories.

selves than with other atoms ; the molecules of these gases are thus probably monatomic. And as a matter of fact the specific heat c for each of these gases is found to be exactly equal to 3, at all temperatures (experiments were carried out up to 2,500° C. with argon).

In short, when molecules are monatomic, they will not be caused to spin when they strike each other even at speeds of the order of a kilometre per second. In this respect the atoms behave as though they were perfectly rigid, smooth spheres (Boltzmann). But this is only one possibility, and all that is suggested by absence of rotation is that two atoms approaching one another are repelled by a force directed towards the centre of gravity of each of the atoms, which therefore cannot be caused to spin. In the same way (with the difference that attractive forces are operative) a comet that is strongly deviated by its passage close to the sun does not communicate any rotation to the latter.

In other words, at the instant when two atoms rushing towards each other undergo the sudden change in velocity that constitutes an impact, they affect each other as though they were two point *centres of repulsion*, of dimensions infinitely small by comparison with their distance apart.

In fact, we shall ultimately (para. 94) come to the conclusion that the material part of the atom is probably enclosed within a sphere, of extremely small diameter, which repels with great violence all other atoms that approach within a certain limiting distance, so that the minimum distance between the centres of two atoms moving towards each other with velocities of the order of a kilometre per second lies well above the real atomic diameter. In the same way, the range of the guns on a battleship very greatly exceeds the circumference of the ship itself. This minimum distance is the radius of a *sphere of protection* which is concentric to the atom and vastly greater than it. We shall find that an altogether new phenomenon is produced when we succeed in increasing greatly the speeds that precede impact, and that then the atoms pierce the protecting spheres instead of rebounding from them (para. 113).

MOLECULAR AGITATION 67

41.—A Serious Difficulty.—Even if the material part of the atom is concentrated within a sphere very small relative to the distance from it at which impact takes place, it appears impossible to suppose that its symmetry can both be and remain such that, at the moment of impact, the repelling forces should always be accurately directed along the line joining the atomic centres. Now, contrary to what might be expected from a superficial inquiry, this is a case where a very close approximation is not sufficient and we have therein an exception to the general rule of very great interest. *However few in number* the atoms may be that deviate from the standard condition of symmetry, they will end by gaining rotational energy equal to their energy of translation. And it is easy to see that the more difficult it is to impart rotation by impact, the more difficult will it be to check rotation already acquired, so that only the time taken to reach statistical equilibrium between the two energies will be affected, and not the ratio between them once equilibrium has been reached. Boltzmann has laid stress upon this point and has raised the question whether the time taken to reach equilibrium might not be considerable in comparison with the duration of our measurements.

But this is quite inadmissible, for whether conducted very rapidly (during an explosion) or of long duration, such measurements always give the same values for the specific heat of argon, for example. We are thus faced with a *fundamental* difficulty. It can be removed, but only by postulating a new and rather peculiar property of matter.

42.—Rotational Energy of the Polyatomic Molecules.—It is now natural to inquire what the value of the specific heat c will be when the molecules can cause each other to spin on striking.

Boltzmann has succeeded, without making any fresh hypothesis, in generalising the results of the statistical calculations by which he established the equality between the mean translational energies of the molecules. He has thus been able to calculate what, under standard conditions of molecular agitation, the ratio between the mean translational and rotational energies should be, for a given molecule, if that molecule may be regarded as a solid body.[1]

In the general case where this solid body has no axis of revolution, a very simple result is obtained and it is found that the two kinds of energy are equal. Increase in rotation will therefore absorb 3 calories per degree, the same as the translational increase, which will make 6 calories in all (or, more exactly, 5·96) for the mo ecular heat c.[2]

But if the molecule is composed, dumb-bell like, of two atoms only, each separately comparable to a perfectly smooth sphere (or better, as we have seen, to mutually repulsive centres of force), no kind of impact can impart rotation to the atoms about the axis of revolution joining the centres of the spheres, and Boltzmann's statistical calculation shows that in that case the mean energy of rotation of the molecule will be $\frac{2}{3}$ only of the mean translational energy. Rotational energy will then absorb 2 calories per degree, the translational energy absorbing 3, making 5 in all (or more accurately 4·97) for the heat c.

If finally the molecule is not solid, any deformation or internal vibration caused by impact will absorb still more energy, and the specific heat will rise above 5 calories if the molecule is diatomic and above 6 if it is polyatomic. On the whole, these results agree with the experimental data.

To begin with, for a large number of diatomic gases the specific heat c has sensibly the same value, equal to 5 calories, as is demanded for molecules that may be regarded as smooth, rigid dumb-bells. This is the case (the measurements having been made at about the ordinary temperatures) for oxygen O_2, nitrogen N_2, hydrogen H_2, hydro-

---

[1] It may be remembered that in stereo-chemistry (para. 24) at least approximate rigidity is attributed to the molecule.

[2] In other terms (which are often used):—The condition of a molecule is defined, from the point of view of energy, by the three components along three fixed axes of the speed of translation and by the three components of the speed of rotation. These six components, which can be independently chosen, therefore represent *six degrees of freedom*. For every rise in temperature of 1°, and considering 1 gramme molecule, the energy corresponding to each component will take up 1 calorie : *energy is equally divided between the degrees of freedom.* (For a rigid, smooth, spinning diatomic molecule, two only of the components of the rotational energy are independent, and the number of degrees of freedom falls to five.)

chloric acid HCl, carbon monoxide CO, nitric oxide, NO, etc.

For other diatomic gases (iodine I_2, bromine Br_2, chlorine Cl_2, iodine monochloride ICl) the heat c is from 6 to 6·5 calories. Now these gases happen to be those that split up into monatomic molecules at temperatures that we are able to reach (in the case of iodine dissociation is already complete at about 1,500°). We are perhaps justified in supposing that this dissociation is preceded by some internal modification of the molecule and that the union between the atoms is slackened, energy being absorbed, before the final rupture occurs.

Finally, for polyatomic gases, we should expect, with Boltzmann, that the heat c would be equal or greater than 6 calories. And such indeed is found to be the case with the values obtained for water vapour and methane. More often the number found is considerably greater (8 for acetylene, 10 for carbon bisulphide, 15 for chloroform, 30 for ether). Since the probability of internal modification as a result of impact might be expected to be higher the more complex the molecule, these high values are not to be wondered at.

43.—THE INTERNAL ENERGY OF THE MOLECULE CAN VARY ONLY IN DISCONTINUOUS STEPS. — The various monatomic gases (such as mercury or argon) have shown us that the internal energy of the atoms does not depend upon the temperature. We may reasonably suppose, therefore, that the internal energy absorbed by a polyatomic molecule reappears entirely in the form of oscillations of the unchanged atoms that make up the molecule about their positions of equilibrium, which means that at any instant the moving atoms possess both kinetic and potential energy, due to their oscillation.

It is very remarkable that we cannot take the energy of this oscillation as having a continuous value, capable of variation by insensible degrees. If we could do so, Boltzmann's statistical argument could be extended to the case of vibrating atoms, and, considering only diatomic molecules, the increment of heat absorbed as kinetic energy of oscillation would be $\frac{R}{2}$, or 1 calorie, for each rise of 1° besides the heat absorbed as mean potential energy of oscillation.[1]

The specific heat c of a diatomic gas, probably equal to 7, could not therefore in any case lie between 5 and 6 and, more simply still, would never be lower than 6, for oscillation of continuously varying amplitude would begin to appear only above a certain temperature.

Now we have seen that this is not the case. The specific heat of the diatomic gases is generally about 5 ; it increases slowly with rise in temperature. Thus its value (Nernst) for oxygen is 5·17 at 300°, 5·35 at 500°, and 6 at 2,000°, at which temperature oxygen behaves like chlorine or iodine in the neighbourhood of the ordinary temperatures.

The above values for the specific heat, which are in all cases lower than those demanded by the very natural hypothesis of a continuously variable internal energy of oscillation, are explicable if certain molecules, increasing progressively in number, become modified in a *discontinuous* fashion as the temperature rises.

Since these low values are always met with as the molecule approaches the point of dissociation into atoms (first iodine, bromine and chlorine ; then oxygen, nitrogen and hydrogen), it is reasonable to suppose that these discontinuities accompany the sudden loosening of the valencies that bind the atoms together, each diminution in solidity absorbing a definite *quantum* of energy. Similarly, when we wind up a clock, we can feel through the fingers the energy stored in the spring increasing by indivisible quanta.

We are therefore left with the probable conclusion that the energy in each quantum is stored within the molecule in the form of oscillatory energy ; but we must suppose, contrary to our experience of vibrating systems on the usual

---

[1] This second increment is also found to be 1 calorie, if, as in the pendulum, the force urging each atom towards its position of equilibrium is proportional to its *elongation* (or distance from the equilibrium position), in which case, as with the pendulum, the mean potential and kinetic energies of oscillation would be equal (this is an extension of the theorem indicated in the note to para. 42, on the equipartition of energy).

dimensional scale, that the internal oscillatory energy of a molecule can vary only by discontinuous steps. Though at first sight discontinuity of this kind may seem strange, we are as a matter of fact prepared to support the assumption in view of Einstein's brilliant extension of the hypothesis that enabled Planck to explain the mechanism of isothermal radiation, as we shall see later (para. 88).

According to this hypothesis, the energy of each oscillator varies by equal quanta. Each of these quanta, each of these specks of energy, is moreover the product $h\nu$ of the frequency ν (the number of vibrations per second) peculiar to the oscillator, by a universal constant h, independent of the nature of the oscillator.

Having once granted this, it is possible, as Einstein showed, by making certain simple hypotheses as to the probable distribution of energy between the oscillators, to calculate the specific heat at any temperature as a function of the frequency ν. When the frequency is sufficiently small or the temperature sufficiently high, we find, as on Boltzmann's theory, that the energy is equally divided between the degrees of freedom corresponding to translation, rotation, and oscillation.

44.—MOLECULES IN A STATE OF CONSTANT IMPACT : SPECIFIC HEAT OF SOLID BODIES.—Up to the present I have not considered the potential energy developed at the actual moment of impact—when, for instance, two molecules approach each other with equal velocities and come to rest the one against the other before rebounding with their velocities reversed. For each molecule, the potential energy of impact is in the mean zero in a gas where the duration of impact is very small by comparison with the time that elapses between two impacts ; in other words, at any instant chosen at random, the potential energy of impact of a molecule is in general non-existent and its mean value is nothing. This commonsense argument, which I owe to M. Bauer, is sufficient to show, without calculation, that the principle of equipartition of energy cannot be extended to the case of energy of impact.

But if the gas is progressively compressed, impacts will

become more and more numerous and the fraction of the total energy present at each instant in the form of potential energy of impact will continuously increase. After a certain compression has been reached, the condition of the gas will be such that practically no single molecule can be regarded as free.

Though direct evidence is lacking, it is possible that the molecule may then be much less rigid than a gaseous molecule composed of two or three atoms, because each atom will be attracted towards neighbouring atoms outside the molecule by cohesive forces comparable in magnitude with those which urge them towards the other atoms in the molecule. This brings us to the conclusion that each atom is readily displaceable in all directions about a certain mean equilibrium position.

The laws of elasticity for solids (reaction proportional to deformation) leads to the supposition that the force urging the atom back towards its position of equilibrium is proportional to its displacement, which means that the atomic vibrations are harmonic and that in the mean the potential energy is equal to the energy of motion.

Finally, assuming, as was done in Boltzmann's statistical calculations, that molecular agitation is a permanent condition, and considering a solid in thermal equilibrium with a gas, we shall find that the mean kinetic energy has the same value for each atom of the solid and each molecule of the gas. On raising the temperature by 1°, each gramme atom of the solid body absorbs 3 calories due to increase in energy of motion of the component atoms, and, according to what has been said as to the equality between the kinetic and potential energies, it also absorbs 3 calories due to increase in the potential energies of these atoms. This makes 6 calories in all, and we obtain Dulong and Petit's Law (para. 15).

But this gives us no explanation of why the specific heat of solids tends to zero at very low temperatures, Dulong and Petit's law becoming quite inaccurate. As we shall see later (para. 90), Einstein has succeeded in explaining this variation of specific heat with temperature but only by assuming (as he had done for the internal oscillations of gaseous molecules) that the actual energy of oscillation of each atom varies by indivisible quanta, of the form $h\nu$, greater or less according as the frequency ν of the oscillation possible for the atom is high or low.

45.—GASES AT VERY LOW TEMPERATURES : EVEN ROTATIONAL ENERGY VARIES DISCONTINUOUSLY.—At very low temperatures peculiarities, at first sight hard to explain, are observed with gases as well as with solids.

Even at the temperature of melting ice (273° absolute) the specific heat of hydrogen is only 4·75, and is thus distinctly lower than the theoretical value 4·97. The discrepancy is not great, but, as Nernst has justly pointed out, it lies in the direction absolutely irreconcilable with Boltzmann's results on rotational energy. Under his direction investigations have been carried out by Eucken at a very low temperature, and have led to the surprising result that, below 50° absolute, the specific heat of hydrogen becomes 3, as with the monatomic gases ! For other gases the specific heat at low temperatures also falls below the theoretical value (though at much lower temperatures than hydrogen), and in fact it seems probable that at sufficiently low [1] temperatures all gases have the same specific heat as the monatomic gases, namely 3 ; that is to say, the molecules, although not spherical, no longer by their impacts impart to each other rotational energy comparable with their energy of translation.

This is incomprehensible, after what has been said above, if the rotational energy can vary by insensible degrees. And we are therefore *forced* to conclude, with Nernst, that this rotational energy does indeed vary by indivisible quanta like the atomic oscillation within the molecule. We may express this result by stating that the angular velocity of rotation varies in a discontinuous manner. This is indeed strange, but if we bear in mind that we are dealing, as we shall see later (p. 160), with rotations so rapid that each molecule revolves more than a million times in one millionth of a second,[2] we need not be surprised at the possibility of other

---

[1] Liquefaction can always be avoided by working under reduced pressure.
[2] Which means that the acceleration must have a colossal value.

properties of matter becoming manifest, which are quite imperceptible when looked for in rotating systems of the kind to which we are accustomed.

Coming back, therefore, to the case of monatomic molecules, we begin to suspect the solution to the problem that at first sight seemed so perplexing. If two of these atoms are not caused to spin when they strike each other, although they are not mutually repelled by forces acting *exactly* between their centres, the cause is certainly to be sought in a very marked discontinuity in the energy of rotation. Obliged to spin rapidly or not at all, they would in general be able to acquire the high minimum rotational energy by impact only at very high temperatures, and it has not been possible up to the present to measure specific heats at such temperatures. This idea will be developed later (para. 94), where it will be shown that the atom in reality occupies but little of the space at the centre of its sphere of protection.

Molecular Free Paths.

46.—The Viscosity of Gases.—Although molecules move with velocities of several hundreds of metres per second, even gases mix but slowly by diffusion. This can be explained if we remember that each molecule, being continually driven in all directions by the impacts it receives, may take a considerable time to move from its original position.

Thus bearing in mind the way in which the movements of a molecule are obstructed by neighbouring molecules, we are led to the conception of the *mean free path* of a molecule, which is the mean value of the path traversed in a straight line by a molecule between two successive impacts. It has been found possible to calculate this mean free path (and we shall find a knowledge of it of service later on in calculating the size of molecules) by establishing its connection with the viscosity of gases.

We are certainly not accustomed in practice to regard gases as viscous substances. As a matter of fact, they are very much less viscous than liquids, but their viscosity is measureable nevertheless. Thus, suppose we have a well-

MOLECULAR AGITATION

polished horizontal disc, placed in a gas, and revolving with a uniform motion about a vertical axis passing through its centre. It will not merely slip round in the layer of gas in immediate contact with it but will carry the layer round too. The first layer will then carry round with it, owing to its friction, an adjacent layer, and so on, until gradually the movement will be transmitted throughout the gas by "internal friction," just as in a liquid ; consequently a second disc parallel to the first and suspended above it by a torsion thread will ultimately be carried round by the tangential forces thus transmitted, until the torsion balances them (which makes it possible to measure them).

The phenomenon is easily explained by the molecular agitation hypothesis. To make this clear, let us first imagine two train loads of travellers moving in the same direction along parallel sets of rails and at nearly equal speeds. We may imagine these travellers amusing themselves by constantly leaping from one train to the other, alighting with a slight impact at each leap. As a result of these impacts the travellers alighting on the slower train would slowly increase its speed, and on the other hand would diminish the speed of the faster train when they leaped upon it. The two speeds would thus ultimately become equal, just as if they had been equalised by direct friction ; indeed the process is actually a frictional effect, with a mechanism that we are able to perceive.

The same effect will be produced when two gaseous layers slide the one upon the other. We may express this condition by supposing that the molecules in, say, the lower layer have, on the average, a certain excess of velocity, in a fixed horizontal direction over the molecules in the upper layer. But the molecules are moving in all directions, and in consequence they will continually be projected from the lower into the upper layer. They will carry with them their excess speed, which will soon be distributed among the molecules in the upper layer, thus increasing slightly its velocity in the given direction ; at the same time, as a result of the action of molecules projected from the upper layer, the speed of the lower layer will diminish slightly. Equalisation of the two speeds will therefore ensue, unless, of course, their constant difference is artificially maintained by some external means.

The effect of a molecular projectile on a layer will be the greater the farther off the layer is from which it comes, for then it must necessarily bring with it a larger excess of speed ; this will occur the more often the greater the mean free path. Furthermore, the effect of the bombardment, for the same free path, must be proportional to the number of projectiles that a layer receives from others in its vicinity. We are therefore prepared to accept the results of the more detailed [1] mathematical analysis by which Maxwell showed that the coefficient of viscosity ζ (or tangential force per square centimetre for a velocity gradient equal to 1) should be very nearly equal to one-third of the product of the following three quantities : d the gas density, G the mean molecular velocity, and L the mean free path :—

$$\zeta = \frac{1}{3} \text{G} . \text{L} . d.$$

It is fairly obvious that for a density, say, 3 times less the free path will be 3 times greater. If, therefore, L varies inversely with d, the product G . L . d is constant ; *the viscosity is independent of the pressure* (at a given temperature). This law appeared very remarkable when it was first announced, and its verification (Maxwell, 1866) constituted one of the first important successes of the kinetic theory.[2]

Since the viscosity is measurable [3] (a method for so doing has been indicated), it appears that all the quantities in Maxwell's equation are known except the free path L, which can therefore be calculated. For oxygen or nitrogen (under normal conditions) the mean free path is very nearly

---

[1] The reasoning is very similar to that which gives gas pressure in terms of the molecular velocity.

[2] Under very low pressures care must be taken that the dimensions of the measuring apparatus (such as the distance between the plates that are caused to rotate by the internal friction of the gas) are sufficiently large by comparison with the free path ; otherwise the theory is inapplicable.

[3] Order of magnitude : ·00018 dyne for oxygen (under normal conditions). Water at 20° C. is about 50 times more viscous.

one ten-thousandth of a millimetre ($\cdot 1\mu$). For hydrogen it is very nearly double that value. Under the very low pressures reached in Crookes' tubes, it often happens that a molecule may move several centimetres in a straight line without meeting another molecule.

During one second a molecule describes as many free paths as it receives impacts, and its total path traversed in the same time should be the mean speed G ; *the number of impacts per second* is therefore the quotient of that speed by the mean free path. This gives a total of very nearly five thousand million for air molecules under normal conditions.

47.—THE MOLECULAR DIAMETER, AS DEFINED BY IMPACT. —The mean free path has been calculated from a knowledge of its relation to the viscosity of gases. It can also be deduced from the simple postulate that the free paths must be the greater the smaller the molecules (they would never strike each other at all, if they were points without magnitude).

Clausius was of the opinion that molecules might be regarded, without great error, as spherical balls having a diameter equal to the distance between the centres of two molecules at the moment of impact. This condition of sphericity might be expected to hold approximately for monatomic molecules. It must be borne in mind, as has been pointed out above, that the distance between the centres at the moment of impact (probably slightly variable according to the violence of the impact) is equal to the radius of a *sphere of protection* maintained by the intense forces of repulsion, and is not necessarily equal to the diameter of the material portion of the molecule. Several difficulties in connection with the kinetic theory arise solely from the fact that the same expression " molecular diameter " is used to denote magnitudes that may be widely different.[1] To avoid all confusion we shall call the quantity that Clausius calls the molecular diameter the *diameter of impact* or *radius of protection*. When two molecules strike against each other their *spheres of impact* are tangential.

---

[1] Diameter of the actual molecular mass, diameter of impact, diameter as defined by the state of the molecules when brought close together in the solid state and when cold, the diameter of the conducting sphere having the same effect as the molecule, etc.

With these reservations, let the volume occupied by a gramme molecule of a gas be v, so that there are $\dfrac{N}{v}$ molecules in unit volume, moving with mean velocity G. Suppose that at a given moment all the molecules become fixed in their positions, with the exception of one that retains the velocity G and rebounds from molecule to molecule with a mean free path L' (which differs, as we shall see, from the free path L that obtains when all the molecules are in motion). Consider the series of cylinders of revolution having as axes the successive directions of the moving molecule and a circle of radius D as base, D being the distance we have just defined as the diameter of impact ; the mean volume of these cylinders is $\pi D^2 L'$. After a large number of impacts, say p, the total volume of the whole series of cylinders, which is equal to $p\,\pi D^2 L'$, will include just as many of the fixed molecules as there are separate cylinders. Since unit volume contains $\dfrac{N}{v}$ molecules, we have :—

$$\frac{N}{v} p \pi D^2 L' = p, \text{ or } N\pi D^2 = \frac{v}{L'}.$$

Clausius was satisfied with this equation, in which he inadvertently assumed equality between L and L'. Maxwell pointed out that the chances of impact are greater for a molecule moving with a mean speed G when the other molecules are in motion also ; for then the speed of two molecules with respect to each other [1] takes the higher mean value of $G\sqrt{2}$. From this it follows that L' must be equal to $\sqrt{2}\,.\,L$.

In short, Clausius' calculation, corrected by Maxwell, gives the *total surface of the spheres of impact* of the N molecules in a gramme molecule, according to the equation

$$\pi N D^2 = \frac{v}{L\sqrt{2}}$$

where L stands for the free path when the volume of the

---

[1] Let R be a relative velocity, the resultant of the velocities u and u', θ being the angle between their directions ; we then have for R^2 the value $(u^2 + u'^2 - 2u\,.\,u'\cos\theta)$ or, in the mean, the value $2U^2$.

gaseous gramme molecule is v; this free path can be deduced from the viscosity of the gas.

Applying this equation in the case of oxygen (v equal to 22,400 cubic centimetres and L equal to $\cdot 1\mu$), we find that the spheres of impact of the molecules in 1 gramme molecule (32 grammes) have a total surface of 16 hectares; placed side by side in the same plane, they would cover an enormous surface; slightly more, in fact, than 5 hectares.

A further relation between Avogadro's number N and the diameter D of the sphere of impact would give us these two magnitudes.

In the first place, it may be pointed out that the diameter D, determined when molecular impact is violent, is probably a little less than the distance to within which the centres of molecules approach when the body under consideration is liquid (or vitreous) and as cold as possible. Moreover, in a liquid, the molecules cannot be more closely packed together than are the shot in a pile of shot. *The total volume of the spheres of impact* (the volume $N \cdot \frac{\pi D^3}{6}$ of the spheres of protection) is consequently less than $\frac{3}{4}$ of the limiting volume reached by the gramme molecule when liquefied or solidified at very low temperatures, and this limiting volume is known. The rough relationship thus obtained, combined with the exact expression for the surface ($N \pi D^2$), leads to values too high for the diameter D and too low for Avogadro's number N.

The calculation for mercury (which is monatomic) gives the diameter of impact for mercury atoms as less than one-millionth of a millimetre and for Avogadro's number a value above 44×10^{22}.

48.—VAN DER WAAL'S EQUATION.—As a matter of fact, the upper limit thus set to the size of the molecules must be fairly near to the actual value, as may be shown by the line of reasoning employed by van der Waals, of whose work I wish to give some account.

We know that fluids obey the gas laws only when beyond a certain degree of rarefaction (oxygen under a pressure of 500 atmospheres does not obey Boyle's law at all). The fact is that under such conditions certain influences, which are negligible in the gaseous state, become of great importance. In order to derive the law of compressibility for condensed fluids, it is, in the opinion of van der Waals, only necessary to correct the theory as applied to gases on the two points following :—

In the first place, in calculating the pressure due to molecular impact, it is assumed that the volume of the molecules (more accurately, the volume of the spheres of impact) is negligible compared with the volume of the space in which they move. Van der Waals, taking this circumstance into account, obtained by a more complete analysis the equation

$$p(v - 4B) = RT,$$

where B stands for the volume of the spheres of impact of the N molecules in a gramme molecule occupying the volume v under a pressure p at an absolute temperature T. This equation, however, only has the above simple form if B, without being negligible, is nevertheless small compared with v (we may take it that it must be less than one-twelfth of v).

In the second place, the molecules in a fluid attract one another, and this diminishes the pressure that the fluid would exert if its cohesion were *nil*. Taking this second circumstance into account, a simple calculation gives the following equation, which is applicable to fluids in general :—

$$\left(p + \frac{a}{v^2}\right)(v - 4B) = RT,$$

where a is a factor expressing the fluid's cohesion, which exerts its influence in proportion to the square of the density. This is *van der Waals' equation*.[1]

This well-known equation agrees sufficiently well with experiment so long as the fluid is not too condensed (it holds roughly even for the liquid state). In other words, for every fluid two numbers can be found which, substituted for a and B, render the equation very nearly exact for all corresponding values of p, v, and T. (The two values for a and B can be determined by assuming that the equation

---

[1] It is more usual to write b instead of 4B.

holds accurately for the fluid under two given sets of conditions and thus obtaining two equations in a and B.)

Once B is known, we can get the *surface of impact* and the *volume of impact* of the N molecules in a gramme molecule from the equations

$$\pi N D^2 = \frac{v}{L\sqrt{2}}$$

$$\frac{\pi N D^3}{6} = B,$$

which will give us all the magnitudes we are seeking (1873).

49.—MOLECULAR MAGNITUDES. — N has been worked out for oxygen and nitrogen, a value very nearly equal to 45×10^{22} being obtained, (to be precise, taking the diameters to be about 3×10^{-8}, we get 40×10^{22} for oxygen, 45×10^{22} for nitrogen, 50×10^{22} for carbon monoxide, a degree of concordance sufficiently remarkable). The substances chosen are not those best suited to the calculation, since we are forced to calculate the "diameters" of molecules that are certainly not spheres. A monatomic substance only, such as argon, can give a trustworthy result. Employing the data available for this substance, it is found that the volume B of the spheres of impact, for 1 gramme molecule (40 grammes), is 7·5 cubic centimetres. This leads to a diameter of impact for the molecule equal to $2·85 \times 10^{-8}$, so that

$$D = \frac{2·85}{100,000,000} \text{ centimetres.}$$

and to a value for N equal to 62×10^{22}, or

$$N = 620,000,000,000,000,000,000,000.$$

The mass of any atom or molecule whatever follows. For instance, the mass of the oxygen molecule will be $\frac{32}{N}$, or 52×10^{-24}; similarly the mass of the hydrogen atom will be $1·6 \times 10^{-24}$, or

$$\frac{1·6}{1,000,000,000,000,000,000,000,000} \text{ gramme.}$$

Such an atom would be lost in our body almost as completely as our body would be lost in the sun.

The energy of motion $\frac{3}{2} \cdot \frac{R}{N}$. T of a molecule at the temperature 273° A. of melting ice will be $\cdot 55 \times 10^{-13}$ ergs; in other words, the work developed by the stoppage of a molecule would be sufficient to raise a spherical drop of water 1μ in diameter to a height of nearly 1μ.

Finally, the atom of electricity (30), which is the quotient $\frac{F}{N}$ of a faraday by Avogadro's number, will have the value $4\cdot 7 \times 10^{-10}$ (C. G. S. *electrostatic* units), or, if it be preferred, $1\cdot 6 \times 10^{-20}$ coulombs. This is very nearly the one thousand-millionth of the quantity that can be detected by a good electroscope.

The probable error, for all these numbers, is roughly 30 per cent., owing to the approximations made in the calculations that lead to the Clausius-Maxwell and van der Waals equations.

In short, each molecule of the air we breathe is moving with the velocity of a rifle bullet; travels in a straight line between two impacts for a distance of nearly one ten-thousandth of a millimetre; is deflected from its course 5,000,000,000 times per second, and would be able, if stopped, to raise a particle of dust just visible under the microscope by its own height. There are thirty milliard milliard molecules in a cubic centimetre of air, under normal conditions. Three thousand million of them placed side by side in a straight line would be required to make up one millimetre. Twenty thousand million must be gathered together to make up one thousand-millionth of a milligramme.

The Kinetic Theory justly excites our admiration. It fails to carry complete conviction, because of the many hypotheses it involves. If by entirely independent routes we are led to the same values for the molecular magnitudes, we shall certainly find our faith in the theory considerably strengthened.

CHAPTER III

THE BROWNIAN MOVEMENT—EMULSIONS

HISTORY AND GENERAL CHARACTERISTICS.

50.—THE BROWNIAN MOVEMENT.—Direct perception of the molecules in agitation is not possible, for the same reason that the motion of the waves is not noticed by an observer at too great a distance from them. But if a ship comes in sight, he will be able to see that it is rocking, which will enable him to infer the existence of a possibly unsuspected motion of the sea's surface. Now may we not hope, in the case of microscopic particles suspended in a fluid, that the particles may, though large enough to be followed under the microscope, nevertheless be small enough to be noticeably agitated by the molecular impacts?

It is possible that an inquiry on the above lines might have led to the discovery of the extraordinary phenomenon which microscopical observation first brought within our ken and which has given us such a profound insight into the properties of the fluid state.

To our observations on the usual scale, all portions of a liquid in equilibrium appear to be at rest. On placing any denser object in the liquid it sinks, vertically if it is spherical, and we know, of course, that once it has got to the bottom of the containing vessel it will stay there and will not attempt to rise to the surface by itself.

Though these are quite familiar points, they nevertheless are valid only on the dimensional scale to which we are accustomed. We have only to examine under the microscope a collection of small particles suspended in water to notice at once that each one of them, instead of sinking steadily, is quickened by an extremely lively and wholly haphazard movement. Each particle spins hither and thither, rises, sinks, rises again, without ever tending to come to rest. This is the *Brownian movement*, so called after the English botanist Brown, who discovered it in 1827, just after the introduction of the first achromatic objectives.[1]

This remarkable discovery attracted little attention. Those physicists who mentioned the agitation likened it, I think, to the movements of the dust particles to be seen with the naked eye dancing in a sunbeam under the influence of air currents produced by small inequalities in pressure and temperature. But in this case neighbouring particles move in approximately the same direction as the air currents and roughly indicate the conformation of the latter. The Brownian movement, on the other hand, cannot be watched for any length of time without it becoming apparent that the movements of any two particles are completely independent, even when they approach one another to within a distance less than their diameter (Brown, Wiener, Gouy).

The agitation cannot, moreover, be due to vibration of the object glass carrying the drop under observation, for such vibration, when produced expressly, produces *general* currents which can be recognised without hesitation and which can be seen superimposed upon the irregular agitation of the grains. The Brownian movement, again, is produced on a firmly fixed support, at night and in the country, just as clearly as in the daytime, in town and on a table constantly shaken by the passage of heavy vehicles (Gouy). Again, it makes no difference whether great care is taken to ensure uniformity of temperature throughout the drop; all that is gained is the suppression of the general convection currents, which are quite easy to recognise and which have no connection whatever with the irregular agitation under observation (Wiener, Gouy). Great diminution in the intensity of the illuminating light or change in its colour is without effect (Gouy).

Of course, the phenomenon is not confined to suspensions in water, but takes place in all fluids, though more actively

[1] Buffon and Spallanzani knew of the phenomenon but, possibly owing to the lack of good microscopes, they did not grasp its nature and regarded the " dancing particles " as rudimentary animalculæ (Ramsay : Bristol Naturalists' Society, 1881).

THE BROWNIAN MOVEMENT—EMULSIONS

the less viscous the fluid.[1] Thus it is just perceptible in glycerine and extremely active, on the other hand, in gases (Bodoszewski, Zsygmondy).

Incidentally, I have been able to observe it with minute spheres of water supported by the "black spots" on soap bubbles. The spherules were 100 to 1,000 times thicker than the thin film which served to support them. They thus bore to the black spots very nearly the same relationship that an orange bears to a sheet of paper. Their Brownian movement, which is negligible in the direction perpendicular to the pellicule, is very active in the plane of the latter (almost as active as if the spherules were in a gas).

In a given fluid the size of the grains is of great importance, the agitation being the more active the smaller the grains. This property was pointed out by Brown at the time of his original discovery. The nature of the grains appears to exert little influence, if any at all. In the same fluid two grains are agitated to the same degree if they are of the same size, whatever the substance of which they are composed and whatever their density (Jevons, Ramsay, Gouy). Incidentally, the absence of any influence exerted by the nature of the grains destroys any analogy with the displacements of large amplitude undergone by specks of camphor when thrown upon water; the moving fragments moreover finally come to rest (when the water has become saturated with camphor).

In fact—and this is perhaps its strangest and most truly novel feature—the Brownian movement never ceases. Inside a small closed cell (so that evaporation may be avoided) it may be observed over periods of days, months, and years. It is seen in the liquid inclusions that have remained shut up in quartz for thousands of years. *It is eternal and spontaneous.*

All these characteristics force us to conclude, with Wiener

[1] The addition of impurities (such as acids, bases, and salts) has no influence *whatever* on the phenomenon (Gouy, Svedberg). That the contrary has been maintained, after a superficial examination, is due to the fact that impurities cause the small particles to stick to the glass when they happen to touch the sides of the containing vessel; the movement of the remainder, however, is unaffected. We might as well say that the motion of the waves is stopped when we fasten a wave-tossed plank against a quay.

(1863), that "the agitation does not originate either in the particles themselves or in any cause external to the liquid, but must be attributed to internal movements, characteristic of the fluid state," movements which the grains follow more faithfully the smaller they are. *We are thus brought face to face with an essential property of what is called a fluid in equilibrium ; its apparent repose is merely an illusion due to the imperfection of our senses and corresponds in reality to a permanent condition of unco-ordinated agitation.*

This view agrees completely with the requirements of the molecular hypotheses, which indeed find in the Brownian movement such confirmation as was looked for above. Every granule suspended in a fluid is being struck continually by the molecules in its neighbourhood and receives impulses from them that do not in general exactly counterbalance each other ; consequently it is tossed hither and thither in an irregular fashion.

51.—THE BROWNIAN MOVEMENT AND CARNOT'S PRINCIPLE.—We have therefore to deal with an agitation that continues indefinitely and is without external cause. Clearly the agitation cannot go on in contradiction to the principle of the conservation of energy. This condition is satisfied if every increment of velocity acquired by a grain is accompanied by the cooling of the fluid in its immediate neighbourhood, and similarly if every diminution in velocity is accompanied by local heating. It merely becomes apparent that *thermal equilibrium is itself only a statistical equilibrium.*

But it must be remembered (Gouy, 1888) that the Brownian movement, which is a fact beyond dispute, provides an experimental proof of those conclusions (deduced from the molecular agitation hypothesis) by means of which Maxwell, Gibbs, and Boltzmann robbed *Carnot's principle* of its claim to rank as an absolute truth and reduced it to the mere expression of a very high probability.

The principle asserts, as we know, that in a medium in thermal equilibrium no contrivance can exist capable of transforming the calorific energy of the medium into work. Such a machine would, for example, allow of a ship being propelled by the cooling of the sea water ; and because of

THE BROWNIAN MOVEMENT—EMULSIONS

the vastness of such a reserve of energy, this would be of practically the same advantage to us as a machine capable of " perpetual motion." That is to say, it would be doing work without taking anything in exchange and without external compensation. But this *perpetual motion of the second kind* is held to be impossible.

Now we have only to follow, in water in thermal equilibrium, a particle denser than water, to notice that at certain instants it rises spontaneously, thus transforming a part of the heat of the medium into work. If we were no bigger than bacteria, we should be able at such moments to fix the dust particle at the level reached in this way, without going to the trouble of lifting it and to build a house, for instance, without having to pay for the raising of the materials.

But the bulkier the particle to be raised, the smaller is the chance that molecular agitation will raise it to a given height. Imagine a brick weighing a kilogramme suspended in the air by a rope. It must have a Brownian movement, though it will certainly be very feeble. As a matter of fact we shall shortly be in a position to calculate the time we would have to wait before we had an even chance of seeing the brick rise to a second level by virtue of its Brownian movement. (That time [1] will be found to be such that by comparison the duration of geological epochs and perhaps of our universe itself will be quite negligible.) Common sense tells us, of course, that it would be foolish to rely upon the Brownian movement to raise the bricks necessary to build a house. Thus the practical importance of Carnot's principle *for magnitudes and lengths of time on our usual dimensional scale* is not affected ; nevertheless we shall evidently gain a better understanding of the ultimate significance of that law of probability by stating it as follows :—

On the scale of magnitudes that are of practical interest to us, perpetual motion of the second kind is in general so insignificant that it would be foolish to take it into consideration.

It would, moreover, be incorrect to say that Carnot's principle is incompatible with the conception of molecular

---

[1] Considerably more than $10^{10^{10}}$ years.

motions. On the contrary, it follows as a consequence of that motion, though in the form of a law of probability. In order to escape the restrictions imposed by that law and to transform at will all the energy of motion of the molecules in a fluid in thermal equilibrium into work, it must be possible to *co-ordinate*, or to make parallel, the velocities of all of them.

52.—Wiener's researches and conclusions might have exercised a considerable influence on the mechanical theory of heat, then in process of development; but, embarrassed by confused ideas as to the mutual actions of material atoms and "ether atoms," they remained unknown. Sir W. Ramsay (1876), and afterwards Professors Delsaulx and Carbonelle, arrived at a clearer understanding of the manner in which molecular motion is able to produce the Brownian movement. According to them, "the internal movements which constitute the heat content of fluids is well able to account for the facts." And, going more into detail, "in the case of large surfaces, molecular impacts, which cause pressure, will produce no displacement of the suspended body, because taken altogether they tend to urge the body in all directions at once. But, if the surface is smaller than the area necessary to ensure that all irregular motions will be compensated, we must expect pressures that are unequal and continually shifting from point to point. These pressures will not be made uniform by the law of aggregates and, their resultant being no longer zero, they will vary continuously in intensity and direction. . . ." (Delsaulx and Carbonelle).

The same conclusion was reached by Gouy, whose exposition of the question was particularly brilliant (1888), by Siedentopf (1900), and finally by Einstein (1905), who succeeded in formulating a quantitative theory of the phenomenon; I shall give an account of his work later.

However seductive the hypothesis may be that finds the origin of the Brownian movement in the agitation of the molecules, it is nevertheless a hypothesis only. As I shall explain later on, I have attempted (1908) to subject the question to a definite experimental test that will enable us to verify the molecular hypothesis as a whole.

If the agitation of the molecules is really the cause of the Brownian movement, and if that phenomenon constitutes an accessible connecting link between our dimensions and those of the molecules, we might expect to find therein some means for getting at these latter dimensions. This is indeed the case, and we have moreover a choice of methods we may employ. I shall discuss first the one that seems to me the most illuminating.

Statistical Equilibrium in Emulsions.

53.—Extension of the Gas Laws to Dilute Emulsions. —We have seen (para. 26) how the gas laws were extended by van't Hoff to dilute solutions, where *osmotic* pressure (exerted on a *semi-permeable membrane* which stops the passage of the dissolved substance but allows the solvent to pass through) takes the place of pressure in the gaseous state. At the same time (para. 26 : note) we saw that this law of van't Hoff's holds for all solutions that obey Raoult's laws.

Now Raoult's laws are applicable indiscriminately to all molecules, large or small, heavy or light. The sugar molecule, containing as many as 45 atoms, and the quinine sulphate molecule, containing more than 100, exert no greater or less effect than the active water molecule, which contains 3 atoms only.

Is it not conceivable, therefore, that there may be no limit to the size of the atomic assemblages that obey these laws ? Is it not conceivable that even visible particles might still obey them accurately, so that a granule agitated by the Brownian movement would count neither more nor less than an ordinary molecule with respect to the effect of its impact upon a partition that stops it ? In short, is it impossible to suppose that the laws of perfect gases may be applicable even to emulsions composed of visible particles ?

I have sought in this direction for crucial experiments that should provide a solid experimental basis from which to attack or defend the Kinetic Theory. In the following paragraph I shall describe the one that appears to me to be the simplest.

54.—DISTRIBUTION OF EQUILIBRIUM IN A VERTICAL COLUMN OF GAS.—It is well known that the air is more rarefied in the mountains than at sea level and that, in general terms, any vertical column of gas is compressed under its own weight. The rarefaction has been given by Laplàce (who obtained it when working out the connection between altitude and barometric indications).

In order to obtain his law, let us consider a thin horizontal cylindrical element, of unit cross-sectional area and of height h; slightly different pressures p and p' will be exerted on the two faces of the element. There would be no change in the condition of the element if it were to be enclosed between two pistons held in position by pressures equal to p and p'; the difference $(p-p')$ between them must balance the force gm due to gravity which tends to pull the mass m of the element downwards. This mass m, moreover, is to the gramme molecular mass M of the gas as its volume $(1 \times h)$ is to the volume v occupied by the gramme molecule under the same mean pressure, so that

$$p - p' = g \cdot \frac{M}{v} \cdot h.$$

And since the mean pressure differs very little from p, so that we may substitute (from the equation for perfect gases) $\frac{RT}{p}$ for v, we may write

$$p - p' = \frac{M \cdot g \cdot h}{RT} \cdot p$$

or

$$p' = p \left(1 - \frac{M \cdot g \cdot h}{RT}\right).$$

Clearly, when the thickness h of the element is fixed, the ratio between the pressures on its two faces is fixed, whatever the level of the element. For example, in air, at the ordinary temperature, the pressure falls by the same relative amount as we mount each step on a staircase (by about $\frac{1}{40,000}$ of its value if the step is 20 centimetres high). If p_o is the pressure

THE BROWNIAN MOVEMENT—EMULSIONS

at the foot of the stairs, the pressure after mounting the first step is $p_o \left(1 - \frac{M \cdot g \cdot h}{RT}\right)$; it is again lowered in the same ratio after the second step and becomes $p_o \left(1 - \frac{M \cdot g \cdot h}{RT}\right)^2$. After the hundredth step it will be $p_o \left(1 - \frac{M \cdot g \cdot h}{RT}\right)^{100}$ and so on.[1]

Moreover, it does not matter from what level the staircase starts. Hence, since it is clear that when we rise to the same height starting from the same level the fall in pressure does not depend on the number of steps into which we divide that height, it appears that the pressure will fall in the same ratio each time we rise through the height H, no matter from what level we start. In air (at the ordinary temperature) we find that the pressure becomes halved each time we rise through 6 kilometres. (In pure oxygen, at 0° C., 5 kilometres is sufficient to halve the pressure.)

Of course, since the pressure, being proportional to the density, is therefore proportional to the number of molecules in unit volume, the ratio $\frac{p_o}{p}$ between pressures can be replaced by the ratio $\frac{n}{n_o}$ between the numbers of molecules at the two levels considered.

But the elevation required to produce a given rarefaction varies with the nature of the gas. It is apparent from the formula that the ratio between the pressures does not change if the product Mh remains constant. In other words, if the gramme molecular weight of a second gas is 16 times lighter than that of the first, the elevation required to produce the

---

[1] If the staircase had q steps, the ratio $\frac{p}{p_o}$ between the pressures at the top and at the bottom would be

$$\frac{p_o}{p} = \left(1 - \frac{M \cdot g \cdot h}{RT}\right)^q.$$

The calculation is simplified by taking logarithms of the two sides, which gives (using ordinary logarithms to base 10) by a simple transformation

$$2 \cdot 3 \log \frac{p_o}{p} = \frac{M \cdot g \cdot H}{RT},$$

where H is the distance between the higher and lower levels and is regarded as being divided into a very large number q of steps each of height h.

same rarefaction will be 16 times greater in the second gas than in the first. Since it is necessary to rise to a height of 5 kilometres in oxygen at 0° C. before its density is halved, a height 16 times greater (or 80 kilometres) will be necessary in hydrogen at 0° C. to produce the same result.

Below are represented three gigantic vertical gas jars (the largest being 300 kilometres high), containing the same number of molecules of hydrogen, helium, and oxygen

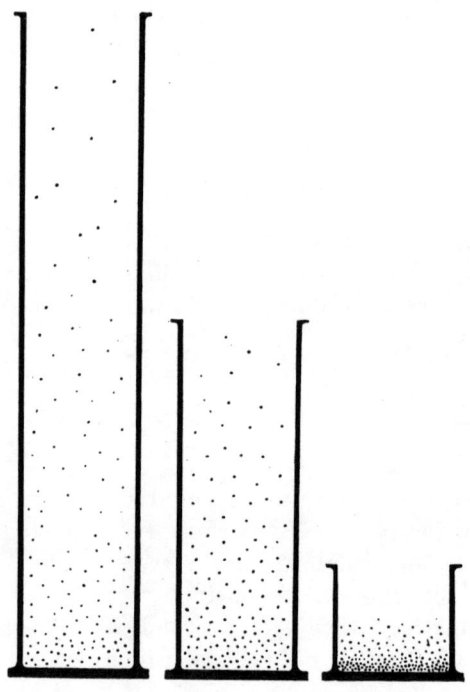

FIG. 3.

respectively. Assuming the temperature to be constant, the molecules will distribute themselves as shown in the figure; the heavier the molecules, the more are they collected together at the bottom.

55.—EXTENSION OF THE THEORY TO EMULSIONS.—The preceding arguments are clearly applicable to emulsions, *if they obey the gas laws.* The particles composing the emulsion

THE BROWNIAN MOVEMENT—EMULSIONS

must be identical, as are the molecules of a gas. The pistons introduced into the argument must be "semi-permeable," stopping the particles but allowing water to pass through. The "gramme molecular weight" of the particles will be Nm, where N is Avogadro's number and m is the mass of a particle. Moreover, the force due to gravity acting on each particle will not be the weight mg of the particle, but its *effective weight*; that is, the excess of its weight over the up-thrust caused by its liquid surroundings. The up-thrust will be equal to $m\dfrac{d}{D}g$, if D is the density of the material of which the particles are composed, and d that of the liquid. A small elevation h will therefore change the concentration of the particles from n to n' according to the equation

$$\frac{n'}{n} = 1 - \frac{N}{RT} \cdot m\left(1 - \frac{d}{D}\right) \cdot gh,$$

which gives at once, as in the case of gases,[1] the degree of rarefaction corresponding to any height H whatever. H may be considered to be subdivided like a flight of stairs into q small steps of height h.

Thus, once equilibrium has been reached between the opposing effects of gravity, which pulls the particles downwards, and of the Brownian movement, which tends to scatter them, equal elevations in the liquid will be accompanied by equal rarefactions. But if we find that we have only to rise $\frac{1}{20}$ of a millimetre, that is, 100,000,000 times less than in oxygen, before the concentration of the particles becomes halved, we must conclude that the effective weight of each particle is 100,000,000 times greater than that of an oxygen molecule. *We shall thus be able to use the weight of the particle, which is measureable, as an intermediary or con-*

---

[1] As with columns of gases, the calculation may be simplified by using logarithms, which gives the following form to the equation for the distribution of the particles:

$$2\cdot 3 \log \frac{n_o}{n} = \frac{N}{RT} \cdot m \cdot \left(1 - \frac{d}{D}\right) g H,$$

or, if we wish to introduce the volume V of a particle:

$$2\cdot 3 \log \frac{n_o}{n} = \frac{N}{RT} \cdot V \cdot (D-d) g H.$$

necting link between masses on our usual scale of magnitude and the masses of the molecules.

56.—THE PREPARATION OF A SUITABLE EMULSION.—My attempts to use the colloidal solutions usually studied (arsenic sulphide, ferric hydroxide, etc.) were unsuccessful. I have, however, been able to use emulsions composed of gamboge and mastic.

Gamboge (which is prepared from a dried vegetable latex) when rubbed with the hand under water (as if it were a piece of soap) slowly dissolves giving a splendid yellow emulsion, which the microscope resolves into a swarm of spherical grains of various sizes. Instead of using the natural grains, it is also possible to treat the gamboge with alcohol, which completely dissolves the yellow matter (which makes up $\frac{4}{5}$ by weight of the crude material). This alcoholic solution, which looks like a bichromate solution, changes abruptly, on the addition of much water, into a yellow emulsion composed of tiny spheres, that appear to be identical with the natural ones.

All resins may be precipitated from alcoholic solution in this way, but often the grains produced are composed of a viscous paste and gradually become stuck together. Out of six other resins tried *mastic* alone appeared suitable. This resin (which gives no natural grains) yields when treated with alcohol a solution that is transformed by the addition of water into a white emulsion, like milk, composed of granules of a colourless, transparent, glassy substance.

57.—FRACTIONAL CENTRIFUGING.—The emulsion having been obtained, it is subjected to an energetic centrifuging (as in the separation of the red corpuscles and serum from blood). The spherules collect together and form a thick sediment; above the sediment is an impure liquid which is decanted. The sediment is treated with distilled water, which brings the grains into suspension once more and the centrifuging process is repeated until the intergranular liquid is practically pure water.

But the purified emulsion contains grains of very various sizes, whereas a *uniform* emulsion (containing grains equal

THE BROWNIAN MOVEMENT—EMULSIONS

in size) is required. The process I use to prepare such emulsions may be likened to fractional distillation. Just as during distillation the fractions evaporating first are richer in volatile constituents, so during the centrifuging of a pure emulsion (made up of grains of the same material) the first layers of sediment formed are richer in large grains, which gives us a means for separating the grains according to size. The technique is easy to imagine and need not be described in detail. I have used rotational speeds of the order of 2,500 revolutions per minute, which produces a centrifugal force 15 centimetres from the axis about 1,000 times that due to gravity. I need scarcely point out that, as in all other kinds of fractionating work, a good separation is a lengthy process. In the most careful of my fractionations I treated 1 kilogramme of gamboge and obtained after several months a fraction containing a few decigrammes of grains having diameters approximately equal to the diameter I wished to obtain.

58.—DENSITY OF THE GRANULAR MATERIAL.—I have determined this in *three* different ways :

(*a*). By the specific gravity bottle method, as for an ordinary insoluble powder. The masses of water and emulsion that fill the same bottle are measured ; then, by desiccation in the oven, the mass of resin suspended in the emulsion is determined. Drying in this way at 110° C. gives a viscous liquid, that undergoes no further loss in weight in the oven and which solidifies at the ordinary temperature into a transparent yellow glass-like substance.

(*b*). By determining the density of this glassy substance, which is probably identical with the material of the grains. This is most readily done by placing a few fragments of it in water, to which is added sufficient potassium bromide to cause the fragments to remain suspended without rising or sinking in the solution. The density of the latter can then be determined.

(*c*). By adding potassium bromide to the emulsion until on energetic centrifuging the grains neither rise nor sink and then determining the density of the liquid obtained.

The three methods give concordant results : for example

96   ATOMS

the same lot of gamboge grains gave the three values 1·1942, 1·194, and 1·195 respectively.

59.—THE VOLUME OF THE GRAINS.—Here again, as with the density, it is possible, on account of the smallness of the grains, to place confidence only in results obtained by several different methods. I have made use of three.

FIG. 4.

A. *Direct measurement of the Radius in the Camera Lucida.* —Considerable error is involved in the measurement of isolated grains (owing to the magnification by diffraction that occurs in the images of small objects). This source of error is very considerably minimised if it is possible to measure the length of a known number of grains in a row. I therefore allowed a drop of *very dilute* emulsion to evaporate

THE BROWNIAN MOVEMENT—EMULSIONS

on an uncovered object-glass. When evaporation is nearly complete, the grains are seen to run together, under the influence of capillary forces, and to collect together into groups a single grain in depth and more or less in rows, in the same way that the shot are arranged in a horizontal section through a pile of shot. It then becomes possible, as can be seen from the photograph reproduced above, to count either the number of grains lying in a row of measured length or the number to be found side by side within a regularly covered area.[1]

At the same time a general check upon the uniformity of the grains sorted out by the operation of centrifuging is obtained. The method gives numbers that are perhaps a little too high (the rows not being quite perfect) ; but owing to its being so direct it cannot be affected by large errors.

B. *Direct Weighing of the Grains.*—In the course of other researches I noticed that, in a feebly acid medium ($\frac{1}{100}$ normal), the grains collect on the walls of the glass without adhering to each other. At any measureable distance from the walls the Brownian movement is not modified. But as soon as a grain chances to reach the wall it becomes fixed, and after a few hours all the grains in a microscopical preparation of known thickness (equal to the distance between the slip and cover-glass) become fixed. It then becomes possible to count at leisure all the grains to be found between the ends of an arbitrary right cylinder (the superficial area of the end being measured in the camera lucida). Further counts are made in specimens taken from various parts of the preparation. Several thousands of grains having been counted in this way, the concentration of the grains is known for a droplet withdrawn, immediately after agitation, from a given emulsion. If the strength of the emulsion is known (by desiccation in the oven) the mass and volume of each grain follows by simple proportion.

C. *Application of Stokes' Law.*—Suppose that, at constant temperature, a tall vertical column of the emulsion under consideration is allowed to stand by itself. Equilibrium

[1] With my best emulsion I have obtained the value $\cdot 373\mu$ for the radius by the first method (from 50 rows of 6 to 7 grains) and $\cdot 369\mu$ by the second (about 2,000 grains distributed over 10^{-5} square centimetres).

distribution will be so far from having been reached that the grains will sink like the minute drops in a mist; we may leave out of account the question of reflux due to the accumulation of grains in the lower layers. The liquid will therefore become gradually clearer in its upper layers. This may readily be observed with an emulsion contained in a capillary tube placed in a thermostat. The edge of the cloud of grains as it sinks will not be very sharply defined, for as a result of the fortuitous fluctuations due to the molecular agitation, the grains will not all fall from the same height; however, by taking the " middle " of the zone, it is possible to evaluate to within nearly $\frac{1}{50}$ the mean value of the distance fallen (it is of the order of a few millimetres per day) and the mean velocity of fall can consequently be obtained.

Furthermore, Stokes has shown (and his conclusions are borne out by experiment, in the case of spheres of directly measurable diameter; 1 millimetre for example) that in a fluid of viscosity ζ the frictional force opposing the motion of a sphere of radius a moving with velocity v is $6\pi\zeta av$. Hence, when the sphere falls with a *uniform* motion under the sole influence of its effective weight, we have

$$6\pi\zeta av = \frac{4}{3}\pi a^3 (D-d)g.$$

Applying this equation to the velocity of descent of the cloud of grains in an emulsion, we have another means for obtaining the radius of the grains (to a degree of accuracy double that attained for the velocity of descent).

The three methods give concordant results, as is shown in the following table, in which the numbers in the same horizontal line give, in microns, the values indicated for the grains in the same emulsion :—

	Rows.	Weight.	Velocity of fall.
I.	·50	—	·49
II.	·46	·46	·45
III.	·371	·3667	·3675
IV.	—	·212	·213
V.	—	·14	·15

THE BROWNIAN MOVEMENT—EMULSIONS 99

Agreement is obtained up to ultra-microscopic magnitudes. The determinations with emulsions III. and IV., which were particularly carefully prepared, show a mean variation of less than 1 per cent. Each of the radii ·3667 and ·212 was obtained by counting about 10,000 grains.

60.—EXTENSION OF STOKES' LAW.—Incidentally, these experiments remove the doubt that had been expressed, with justice (J. Duclaux), as to the propriety of extending Stokes' law to the velocity of falling clouds. Stokes' law expresses the real velocity of a sphere with respect to a fluid, but in the case under consideration it is applied to a *mean* velocity unconnected with the real velocities of the grains; these latter velocities are incomparably greater and are constantly varying.

It cannot now be doubted, in the face of the concordant results given above, that in spite of the Brownian movement the extension of the law is legitimate. But the experiments refer only to liquids.[1] In gases, as I shall show later, Stokes' law ceases to be applicable, not on account of the agitation of the granules, but because the size of the granules becomes comparable with the mean free path of the molecules of the fluid.

61.—METHOD OF OBSERVING AN EMULSION.—Useful observations with the emulsions I have employed cannot be made through heights of several centimetres or even millimetres; heights of less than the tenth of a millimetre only are suitable. Their investigation has therefore been carried out under the microscope. A drop of emulsion is placed in hollow slide (Zeiss hollow slide, having a depth of ·1 millimetre), and is given a plane surface by means of a cover-glass; the edges of the latter are treated with paraffin to prevent evaporation. Two arrangements are possible (Fig. 5).

The preparation may be vertical and the microscope horizontal; it is then possible to see at a single observation the distribution of the emulsion throughout its height. I have made several observations in this way, but no measurements

---

[1] A further condition is necessary (Smoluchowski); the cloud, which must extend to the sides of the tube in which it is sinking (this condition is fulfilled, in our case, in a capillary tube), must not be able to descend as a whole (liquid flowing back up the sides) as such a cloud would do in the atmosphere.

up to the present. Fig. 6 is reproduced from a photograph taken in my laboratory by M. Constantin, using the above arrangement.

The preparation may also be horizontal, with the microscope vertical. The objective used, which is of high power, has a small depth of field, and only those grains in a very thin horizontal section, of the order of a micron in thickness, can be seen sharply defined at any given moment. As the microscope is raised or lowered, the grains in other sectional layers become visible.

Following either procedure it is shown that the distribution of the grains, which is very nearly uniform after the initial disturbance caused by getting the preparation into position

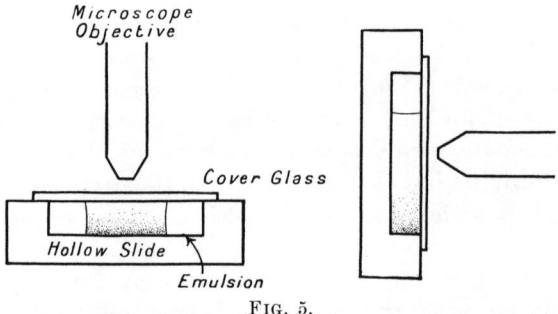

FIG. 5.

has subsided, soon ceases to be so, the lower sections becoming richer in grains; the process of enrichment, however, gradually slackens until a permanent condition is realised in which the concentration diminishes with the height. Fig. 7 was obtained by placing one above the other diagrams showing the distribution of the grains at a given moment at five equidistant levels in a particular emulsion. The analogy between Figs. 6 and 7 and Fig. 3, which represents the distribution of the molecules in a gas, is evident.

The next step is to obtain measurements. We have already the radius a of each grain and its apparent density $(D - d)$, which is the difference between D, the density of the grain and d, the density of water or other intergranular liquid. The vertical distance H between two sections successively examined will be obtained by multiplying the

THE BROWNIAN MOVEMENT—EMULSIONS

vertical displacement H' of the microscope [1] by the index of refraction of the medium separating the slide and coverglass.[2] But we have still to determine the ratio $\frac{n_o}{n}$ between the concentrations of the grains at two different levels.

62.—METHOD OF COUNTING THE GRAINS.—This ratio is obviously equal to the mean ratio between the numbers of grains visible under the microscope at two levels. But the counting of the grains is a difficult matter; when one sees several hundreds of grains, agitated in all directions, continually disappearing and reappearing, it is impossible to estimate their number, even roughly.

FIG. 6.

The simplest procedure is certainly to take instantaneous photographs and then to count at leisure the sharp images of the grains on the plates. But owing to the magnification necessary and the short time available for exposure, an intense light is required, and with grains less than half a micron in diameter I have never succeeded in obtaining good images.

FIG. 7.

I therefore reduced the field of vision by placing in the focal plane a diaphragm consisting of an opaque disc of foil having a very small round hole pierced in it by a needle. The field now visible becomes very restricted and the eye is enabled

[1] Read directly on the graduated head of the micrometer screw of the Zeiss microscope used.
[2] More often I have used water emulsions for these experiments, with a water immersion objective. In that case H is simply equal to H'.

to estimate at once the exact number of grains to be seen at any given moment. The number must be less than 5 or 6. By placing a shutter in the path of the rays that illuminate the preparation they can be allowed to pass at regular intervals, the number of grains perceived on each occasion being noted, thus :—

$$2, 2, 0, 3, 2, 2, 5, 3, 1, 2, \ldots\ldots$$

Starting again at another level, a similar series of numbers will be obtained, such as :—

$$2, 1, 0, 0, 1, 1, 3, 1, 0, 0, \ldots.$$

Owing to the absolute irregularity of the Brownian movement, 200 readings of this kind will clearly be equivalent to one instantaneous photograph embracing a field 200 times as large.[1]

63.—STATISTICAL EQUILIBRIUM IN A COLUMN OF EMULSION. —It is now easy to establish accurately that the distribution of the grains reaches ultimately a permanent condition of dynamic equilibrium. We have only to determine every hour the ratio $\frac{n_o}{n}$ between the concentrations at two fixed levels. This ratio, which is at first nearly 1, increases and tends towards a limit. For a difference in level of ·1 millimetre, with water as the intergranular liquid, the limiting distribution was practically reached after one hour (I have found exactly the same values for $\frac{n_o}{n}$ after three hours and after fifteen days).

The limiting distribution constitutes a *reversible equilibrium*, for if it is displaced, the system returns to its original condition of its own accord. One way of displacing it (*i.e.*, of causing too many grains to accumulate in the lower sections) is to cool the emulsion, which causes an increase in the concentration in the lower layers (I shall return immediately to

[1] By either method uncertainty will arise as to some of the grains observed, which, though barely visible, are sufficiently so for their presence to be guessed at. But such uncertainty affects n_o and n to the same degree. Thus, two different observers, determining $\frac{n_o}{n}$ by means of the spots in a reduced field of vision found the values 10·04 and 10·16 respectively.

THE BROWNIAN MOVEMENT—EMULSIONS

the consideration of this phenomenon), and then allowing it to return to its original temperature ; the distribution then becomes what it was before.

64.—THE LAW ACCORDING TO WHICH THE CONCENTRATION DECREASES.—I have sought to discover whether the distribution of the grains, like that of an atmosphere under the action of gravity, is indeed such that equal elevations are associated with equal rarefactions, so that the concentration falls off in geometric progression.

A series of experiments was carried out with the greatest care, using gamboge grains of radius $\cdot 212\mu$ (using the reduced field of vision method). Cross readings were taken in a cell 100μ deep on four horizontal equidistant planes across the cell at the levels

$$5\mu, 35\mu, 65\mu, 95\mu.$$

The readings gave at these levels, from a count of 13,000 grains, concentrations proportional to the numbers

$$100, 47, 22\cdot 6, 12,$$

which are approximately equal to the numbers

$$100, 48, 23, 11\cdot 1,$$

which are in geometrical progression.

Another series was obtained using larger grains, of mastic (radius $\cdot 52\mu$). Photographs taken at four equidistant levels, one above the other and with 6μ distance between them, show respectively

$$1880, 940, 530, 305$$

images of grains ; these numbers differ but little from

$$1880, 995, 528, 280$$

which decrease in geometrical progression.

In this latter case, the concentration at a height of 96μ would be 60,000 times less than at the bottom. Hence, when permanent equilibrium has been reached, grains will hardly ever be found in the higher layers of such preparations.

Other series might be quoted. In short, as was expected, the rarefaction law is obeyed exactly. But does it lead to these values for the molecular magnitudes that we look for ?

104 ATOMS

65.—A Decisive Proof.—Let us consider grains of such a kind that an elevation of 6μ is sufficient to halve their concentration. To reach the same degree of rarefaction in air, we have seen that a distance of 6 kilometres, which is nearly 10,000 million times as great, is necessary. If our theory is correct, the weight of an air molecule should therefore be one ten thousand-millionth of the weight, in water, of one of the grains. The weight of the hydrogen atom may be obtained in the same way, and it now only remains to be seen whether numbers obtained by this method are the same as those deduced from the kinetic theory.[1]

It was with the liveliest emotion that I found, at the first attempt, the very numbers that had been obtained from the widely different point of view of the kinetic theory. In addition, I have varied widely the conditions of experiment. The volumes of the grains have had values distributed between limits which were to each other as 1 is to 50. I have also varied the nature of the grains (with the aid of M. Dabrowski), using mastic instead of gamboge. I have varied the intergranular liquid (with the help of M. Niels Bjerrum) and studied gamboge grains suspended in glycerine containing 12 per cent. of water, the mixture being 125 times more viscous than water.[2] I have varied the apparent density of the grains, in ratios varying from 1 to 5 ; in glycerine it becomes *negative* (in which case the influence of the changed sign of their weight accumulated the grains in the *upper* layers of the emulsion). Finally, M. Bruhat has, under my direction, studied the influence of temperature and observed the grains first in *super-cooled* water ($-9°$ C.) and then in hot water ($60°$ C.) ; the viscosity in the latter case was half what it was at $20°$ C., so that the viscosity varied in the ratio of 1 to 250.

---

[1] The calculations are simplified if the distribution equation given in the note to para. 55 is used.

[2] The Brownian movement, though much abated, is nevertheless perceptible ; several days are required before a permanent equilibrium is reached. I should have liked to study the distribution in an even more viscous medium, but, when less than 5 per cent. of water was added to the glycerine (very feebly acid), the grains collect upon the sides and permanent equilibrium could no longer be observed. I have subsequently made use of this circumstance in extending the gas laws to these viscous emulsions (para. 78).

THE BROWNIAN MOVEMENT—EMULSIONS

In spite of all these variations, the value found for Avogadro's number N remains approximately constant, varying irregularly between 65×10^{22} and 72×10^{22}. Even if no other information were available as to the molecular magnitudes, *such constant results would justify the very suggestive hypotheses that have guided us*, and we should certainly accept as extremely probable the values obtained with such concordance for the masses of the molecules and atoms.

But the number found agrees with that (62×10^{22}) given by the kinetic theory from the consideration of the viscosity of gases. *Such decisive agreement can leave no doubt as to the origin of the Brownian movement.* To appreciate how particularly striking the agreement is, it must be remembered that before these experiments were carried out we should certainly not have been in a position either to deny that the fall in concentration through the minute height of a few microns would be negligible, in which case an infinitely small value for N would be indicated, or, on the other hand, to assert that all the grains do not ultimately collect in the immediate vicinity of the bottom, which would indicate an infinitely large value for N. It cannot be supposed that, out of the enormous number of values *a priori* possible, values so near to the predicted number have been obtained by chance for every emulsion and under the most varied experimental conditions.

The objective reality of the molecules therefore becomes hard to deny. At the same time, molecular movement has not been made visible. The Brownian movement is a faithful reflection of it, or, better, it is a molecular movement in itself, in the same sense that the infra-red is still light. From the point of view of agitation, there is no distinction between nitrogen molecules and the visible molecules realised in the grains of an emulsion,[1] which have a gramme molecule of the order of 100,000 tons.

Thus, as we might have supposed, an emulsion is actually a miniature ponderable atmosphere; or, rather, it is an atmosphere of colossal molecules, which are actually visible.

[1] Of course, such grains are not *chemical* molecules, in which all the cohesive forces are of the nature of those uniting the carbon to the four hydrogen atoms in methane.

The rarefaction of this atmosphere varies with enormous rapidity, but it may nevertheless be perceived. In a world with such an atmosphere, Alpine heights might be represented by a few microns, in which case individual atmospheric molecules would be as high as hills.

66.—THE INFLUENCE OF TEMPERATURE.—I wish specially to discuss the way in which temperature variation influences the equilibrium distribution; briefly, its effect proves that Gay-Lussac's law applies also to emulsions. We have seen that equilibrium in a column of emulsion, as in a column of gas, is reached between the opposing tendencies due on the one hand to gravity (which urges all the grains in the same direction), and on the other to molecular agitation (which constantly tends to scatter them). The feebler the agitation, that is, the lower the temperature, the more marked will be the subsidence of the column under its own weight.

This *subsidence* when the temperature falls and *expansion* when it rises can be accurately verified without actually causing the temperature to vary very much. This is possible because verification in this case does not necessitate the exact determination, which is always difficult, of the radius of the grains in the emulsion. Let T and T_1 be the temperatures (absolute) of experiment. According to the rarefaction law (note to para. 55) the elevations H and H_1 corresponding in each case to the same rarefaction should be such that

$$\frac{H}{T}\left(1 - \frac{d}{D}\right) = \frac{H_1}{T_1}\left(1 - \frac{d_1}{D_1}\right).$$

(It appears that if the densities do not change, equivalent elevations should be proportional to the inverse ratio between the temperatures.)

M. Bruhat, working in my laboratory, undertook, at my request, to realise the necessary experimental conditions under which verification could be sought, and has succeeded admirably.

The drop of emulsion is placed on the upper surface of a thin, transparent cell in which the temperature is maintained at a fixed value $t°$ C. (measured by a thermo-electric couple) by means of a liquid (hot water or cold alcohol) that flows through it. For cover-glass he used the bottom of a vessel full of liquid (hot water or a non-freezable solution of the same index of refraction as cedar oil) into which he dipped the objective used (water or cedar oil immersion). This liquid was raised to the temperature $t°$ C. (measured by a second thermo-electric couple) by means of a copper tube that traversed it ; a branch stream of the regulating liquid flowed through the tube. Imprisoned in this way the preparation necessarily reaches the temperature $t°$ C.

Counts made under these conditions have verified, to within about 1 per cent., the conclusions reached above, which shows to what degree of exactness the gas laws can be extended to dilute emulsions.[1]

67.—EXACT DETERMINATIONS OF THE MOLECULAR MAGNITUDES.—We have pointed out that the theory of gases, applied to their viscosity, gives the size of the molecules with an approximation of perhaps 30 per cent. Refinements introduced in the actual measurements with gases do not lessen this degree of uncertainty, which is really connected with the simplifying hypotheses introduced in the theory. This is not so in the case of emulsions ; with them the results have the same degree of precision as the experiments upon which they depend. By studying emulsions we are really able to *weigh the atoms* and not merely to estimate their weights approximately.

A series of careful measurements (radius of grain $·212\mu$; number of grains counted at different levels, 13,000) had already given me the value $70·5 \times 10^{22}$ for N. The uniformity of the grains, however, did not appear to me to be sufficiently good. I therefore commenced operations afresh, and a more accurate series (radius $·367\mu$ to within 1 per cent., obtained after prolonged centrifuging ; number of grains counted at various elevations, 17,000) gave for Avogadro's number the probable mean value

$$68·2 \times 10^{22},$$

[1] I have been able to show, with the valuable assistance of M. Constantin, that van der Waal's equation applies to concentrated emulsions (J. Perrin and R. Constantin, C.R. 1914). Constantin has, however, made the unexpected discovery that the constant is negative (the grains, being charged in contact with water, repel instead of attract one another).

from which it follows that the mass of the hydrogen atom is, in grammes,

$$h = \frac{1 \cdot 47}{1,000,000,000,000,000,000,000,000} \; (= 1 \cdot 47 \times 10^{-24}).$$

The other molecular magnitudes follow at once. For instance, molecular energy of translation, which is equal to $\frac{3}{2} \cdot \frac{R}{N} \cdot T$, is very nearly $\cdot 5 \times 10^{-13}$ at the temperature of melting ice.

The atom of electricity will be (in C.G.S. electrostatic units)
$$4 \cdot 25 \times 10^{-10}.$$

The *dimensions* of the molecules, or, more accurately, the diameters of their spheres of impact, can be obtained, now that N is known, from Clausius's equation (para. 48)

$$\pi \, N.D^2 = \frac{v}{L\sqrt{2}}$$

by first calculating the mean free path L for a gramme molecule of the substance occupying the volume v in the gaseous state.

For example, at 370° C. (643° absolute) the mean free path for mercury, under atmospheric pressure (v is equal to $22{,}400 \times \frac{643}{273}$), can be deduced from the viscosity 6×10^{-4} of the gas by means of Maxwell's equation (para. 47), which gives the value $2 \cdot 1 \times 10^{-5}$ for L. This gives $2 \cdot 9 \times 10^{-8}$ (or $\cdot 29$ millimicrons very nearly) for the required diameter.

I have calculated in this way the following diameters :—

Helium	$1 \cdot 7 \times 10^{-8}$
Argon	$2 \cdot 8 \times 10^{-8}$
Mercury	$2 \cdot 9 \times 10^{-8}$
Hydrogen	$2 \cdot 1 \times 10^{-8}$
Oxygen	$2 \cdot 7 \times 10^{-8}$
Nitrogen	$2 \cdot 8 \times 10^{-8}$
Chlorine	$4 \cdot 1 \times 10^{-8}$

These determinations (particularly for the polyatomic molecules), depending as they do upon the definition of protecting spheres, do not carry the same degree of precision that is possible in the case of masses.

CHAPTER IV

THE LAWS OF THE BROWNIAN MOVEMENT

EINSTEIN'S THEORY.

68.—DISPLACEMENT IN A GIVEN TIME.—It is in consequence of the Brownian movement that equilibrium distribution is reached in an emulsion; the more active the movement, the more rapidly does this occur. But the degree of activity, whether high or low, has no influence on the final distribution, which is always the same for grains of the same size and the same apparent density. We have therefore confined ourselves up to the present to the study of the permanent condition of equilibrium, without bothering about the mechanism by which it is reached.

This mechanism has been subjected to a detailed analysis by Einstein, in an admirable series of theoretical papers.[1] The approximate but very suggestive analysis given by Smoluchowski [2] certainly deserves to be mentioned also.

Einstein and Smoluchowski have defined the activity of the Brownian movement in the same way. Previously we had been obliged to determine the " mean velocity of agitation " by following as nearly as possible the path of a grain. Values so obtained were always a few microns per second for grains of the order of a micron.[3]

But such evaluations of the activity are *absolutely wrong*. The trajectories are confused and complicated so often and so rapidly that it is impossible to follow them; the trajectory actually measured is very much simpler and shorter than the real one. Similarly, the apparent mean speed of a grain

[1] Ann. de Phys., Vol. XVII., 1905, p. 549, and Vol. XIX., 1906, p. 371. A complete account of Einstein's theory will be found in my memoir " Les preuves de la realité moléculaire " (Brussels Congress on the Theory of Radiation and Quanta, Gauthier-Villars, 1912).
[2] Bulletin de l'Acad. des Sc. de Cracovie, July, 1906, p. 577.
[3] Incidentally this gives the grains a kinetic energy 100,000 times too small.

during a given time varies *in the wildest way* in magnitude and direction, and does not tend to a limit as the time taken for an observation decreases, as may easily be shown by noting, in the camera lucida, the positions occupied by a grain from minute to minute, and then every five seconds, or, better still, by photographing them every twentieth of a second, as has been done by Victor Henri, Comandon, and de Broglie when kinematographing the movement. It is impossible to fix a tangent, even approximately, at any point on a trajectory, and we are thus reminded of the continuous [1] underived functions of the mathematicians. It would be incorrect to regard such functions as mere mathematical curiosities, since indications are to be found in nature of " underived " as well as " derived " processes.

Neglecting, therefore, the true velocity, which cannot be measured, and disregarding the extremely intricate path followed by a grain during a given time, Einstein and Smoluchowski chose, as the magnitude characteristic of the agitation, the rectilinear segment joining the starting and end points ; in the mean, this line will clearly be longer the more active the agitation. The segment will be the *displacement* of the grain in the time considered. Its projection on to a horizontal plane, as perceived directly in the microscope under ordinary conditions (microscope vertical), will be its *horizontal displacement*.

69.—THE ACTIVITY OF THE BROWNIAN MOVEMENT.—In accordance with the conclusions arrived at from qualitative observation, we shall regard the Brownian movement as *completely irregular* in all directions at right angles to the vertical.[2] This is scarcely a hypothesis ; moreover, we shall verify all its consequences.

This being granted, and without any further hypothesis whatever, it can be proved that the mean displacement of a grain is doubled when the time is increased fourfold ; it becomes tenfold when the time is increased a hundredfold and so on. More precisely, it is proved that the mean

---

[1] *Continuous* because it is not possible to regard the grains as passing from one position to another without cutting any given plane having one of those positions on each side of it.

[2] It is not so in a vertical direction, on account of the weight of the grains.

LAWS OF THE BROWNIAN MOVEMENT

square e^2 of the horizontal displacement during the time t increases in proportion to that time.

The same result holds for half this square or the mean square x^2 of the projection of the horizontal displacement along an arbitrary horizontal axis.[1] In other words, for a given kind of grain (in a given fluid) the quotient $\dfrac{x^2}{t}$ is constant. Clearly greater the more actively the grain is agitated, this quotient characterises *the activity of the Brownian movement* for any particular grain.

It must be borne in mind, however, that this result ceases to be exact when the times become so short that the movement is not absolutely irregular. This must necessarily be so, otherwise the *true* velocity would be infinite. *The minimum time within which irregularity may be expected* is probably of the same order as the time required by a granule, shot into the liquid with a velocity equal to the *true* mean agitational speed, before the frictional effect due to viscosity reduces its initial energy practically to zero. (The same time, moreover, elapses between successive molecular impacts.) We find in this way, for a spherule 1 micron in diameter in water, that the minimum period of irregularity is of the order of the hundred-thousandth of a second. It would be only 100 times greater, or one-thousandth of a second, for a spherule 1 millimetre in diameter, and 100 times smaller for a liquid 100 times more viscous. Lengths of time such as these fall far short of the periods during which it has been possible to observe the movement up to the present.

70.—THE DIFFUSION OF EMULSIONS.—We would expect that, when pure water is left in contact with an aqueous emulsion composed of equal sized grains, *diffusion* of the grains, due to their Brownian movement, would take place into the water by a mechanism quite analogous to that which causes the diffusion, properly so called, of dissolved substances. It is moreover evident that such diffusion should occur the more rapidly the more active the Brownian move-

---

[1] By resolving each displacement along two horizontal axes perpendicular to each other, and applying the theorem as to the square on the hypothenuse and taking the mean, we get at once $e^2 = 2x^2$.

ment of the grains. Making the *single* supposition that the Brownian movement is completely irregular, Einstein's rigorous analysis shows that an emulsion diffuses like a solution,[1] and that the co-efficient of diffusion D is simply equal to half the number that measures the activity of agitation,

$$D = \frac{1}{2} \cdot \frac{x^2}{t}.$$

Again, we are familiar with the idea that, in a vertical column of emulsion, the permanent distribution is main-

[1] Consider a cylinder parallel to Ox, of unit area in cross section, and filled with solution. Suppose that the concentration has the same value at all points in the same transverse section (which will be the case when pure water is carefully superimposed upon a solution of sugar). The loss of dissolved substance I across a section will be, at each instant, the mass of dissolved substance that traverses it in one second from regions of high towards those of low concentration. The fundamental diffusion law states that this loss will be the greater the steeper the fall of concentration across the section :—

$$I = D \cdot \frac{c - c'}{x' - x}.$$

The co-efficient D, which depends on the nature of the dissolved substance, is the co-efficient of diffusion. For instance, taking the case of sugar, the statement that D is equal to $\frac{\cdot 33}{86,400}$ expresses the fact that, for a concentration gradient maintained equal to 1 gramme per centimetre, ·33 gramme of sugar passes across the transverse section considered in one day, or 86,400 times less in a second.

Bearing this in mind, I can indicate a line of reasoning (due also to Einstein) which, although not a rigid proof, is at any rate approximate and which leads to the formula in question.

In a horizontal cylinder, let n' and n'' be the concentrations of the grains in two sections s' and s'' separated by a distance X. The concentration gradient throughout the intermediate section s will be $\frac{n' - n''}{X}$ and a number of grains equal to $D \frac{n' - n''}{X} t$ will traverse the section s during time t. Further, assuming that this result is produced by each grain suffering, during the time t, the displacement X either towards the right or towards the left, we find that $\frac{1}{2} n' X$ traverse s towards s'' and $\frac{1}{2} n'' X$ towards s', which gives, for the total drift towards s'' :—

$$\frac{1}{2}(n' - n'') X.$$

We therefore have

$$\frac{1}{2}(n' - n'') X = D \cdot \frac{n' - n''}{X} ; t,$$

or, better,

$$X^2 = 2 D \cdot t,$$

which is Einstein's equation.

LAWS OF THE BROWNIAN MOVEMENT

tained by the equilibrium between two opposing actions, namely gravity, which constantly drags the grains towards the bottom of the containing vessel, and the Brownian movement, which continually scatters them. We may give precise expression to this conception by stating it in the following form for any given section, the loss by diffusion towards the region of low concentration balances the influx caused by gravity into the regions of high concentration.

In the special case where the grains are spheres of radius a, to which we can attempt to apply Stokes' law (para. 59) (I have shown that the law holds for microscopic spherules (para. 60)), and assuming moreover that at equal concentrations grains or molecules produce the same osmotic pressure, we find that

$$D = \frac{R \cdot T}{N} \cdot \frac{1}{6\pi a \xi},$$

where ξ is the viscosity of the fluid, T its absolute temperature, and N Avogadro's number. Since the coefficient of diffusion is half the activity of the Brownian movement, we can give the equation the equivalent form

$$\frac{x^2}{t} = \frac{R \cdot T}{N} \cdot \frac{1}{3\pi a \xi},$$

in which we can, moreover, replace (para. 35) $\frac{RT}{N}$ by $\frac{2}{3}$ of the mean molecular energy w.

Thus the activity of the agitation (or the rate of diffusion) should be proportional to the molecular energy (or to the absolute temperature), and inversely proportional to the viscosity of the liquid and to the dimensions of the grains.

71.—ROTATIONAL BROWNIAN MOVEMENT.—Up to the present we have considered only changes in the positions of the grains or their *translational* Brownian movement. But it is known that each grain spins in an irregular fashion during its displacement. Einstein has succeeded in establishing an equation, for this *rotational* Brownian movement, comparable with the one given above, for the case of spherules of radius a. If A^2 represents the mean square in time t of the component of the angle of rotation about a given axis, the quotient $\dfrac{A^2}{t}$, which is fixed for a given grain, characterises the *activity* of the rotational Brownian movement and should follow the equation

$$\frac{A^2}{t} = \frac{RT}{N} \cdot \frac{1}{4\pi a^3 \zeta},$$

the activity of the rotational agitation being, as for the translational activity, proportional to the absolute temperature and inversely proportional to the viscosity. It varies, however, inversely with the volume and not inversely with the dimensions of the grain. A sphere of diameter 10 will have translational agitation 10 times and rotational agitation 1,000 times more feeble than a spherule of diameter 1.

It is not possible to indicate here the way in which this equation is derived; we may point out, however, that it implies, for a given granule, *equality between the mean translational and mean rotational energies*, as was predicted by Boltzmann (para. 42). We shall verify this when we succeed in verifying Einstein's equation.

Experimental Verification.

Such, in its broad outlines, is the remarkable theory we owe to Einstein. It is well adapted to accurate experimental verification, *provided we are able to prepare spherules of measureable radius.* Consequently, ever since I became, through M. Langevin, acquainted with the theory, it has been my aim to apply to it the test of experiment. As we shall see, the experiments that I have carried out myself or supervised in others demonstrate its complete accuracy.

72.—The Complicated Nature of the Trajectory of a Granule.—We have assumed that the Brownian movement (at right angles to gravity) is entirely irregular and have seen that this assumption is the basis of Einstein's theory. However probable this may be, it is important that it should be established on an exact basis.

We will deal first of all with the measurement of the successive displacements (horizontal) undergone by the same grain. To accomplish this we have only to note in the camera lucida (under known magnification) the positions occupied by a grain after successive equal time intervals. In the adjoining figure three diagrams are shown, the scale being such that sixteen divisions represent 50 microns. These diagrams were obtained by tracing the horizontal projections of the lines joining consecutive positions occu-

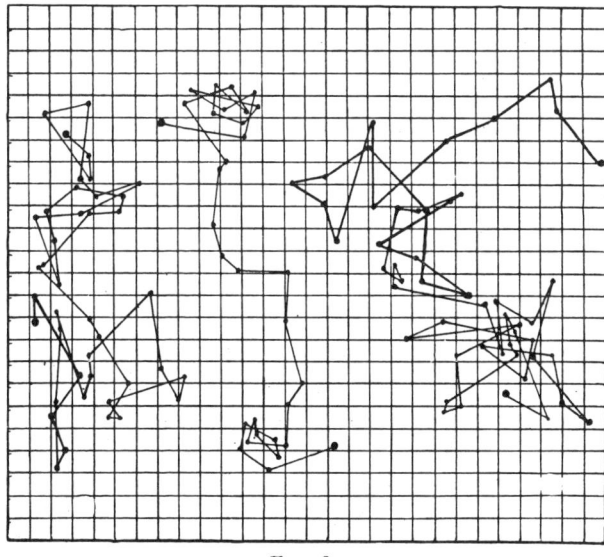

Fig. 8.

pied by the same mastic grain (radius equal to ·53 μ); the positions were marked every 30 seconds. It is clear from these diagrams that the projection of each segment along any horizontal axis whatever can readily be obtained (being given by the abscissæ or ordinates as measured by the squares on the paper).

As a matter of fact diagrams of this sort, and even the next figure, in which a large number of displacements are traced on an arbitrary scale, gives only a very meagre idea of the extraordinary discontinuity of the actual tra-

116   ATOMS

jectory. For if the positions were to be marked at intervals
of time 100 times shorter, each segment would be replaced
by a polygonal contour relatively just as complicated as the
whole figure, and so on. Obviously it becomes meaningless
to speak of a tangent to a trajectory of this kind.

73.—THE COMPLETE IRREGULARITY OF THE AGITATION.—
If the movement is irregular, the mean square X^2 of the

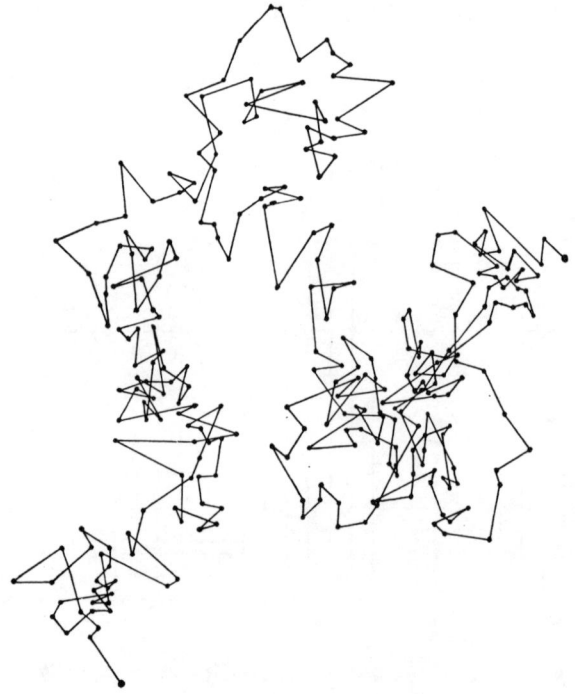

FIG. 9.

projection onto an axis will be proportional to the time.
And as a matter of fact the record of a large number of
positions has shown that this mean square is, for a length of
time of 120 seconds, very nearly twice what it is for 30
seconds.[1]

[1] It is not even necessary to follow the same grain, or to know its size. For
any one series of grains we need only know the displacements d and d' relative
to the lengths of time 1 and 4. The quotient $\frac{d'}{d}$ has the mean value 2.

The possibility of an even more complete verification is suggested by an extension of the line of reasoning developed by Maxwell (para. 35) in connection with *molecular speeds*, to the *displacements* of granules. His arguments should apply equally well in either case.

Thus projections of displacements along any axis, like projections of velocities (considering equal spherules during equal times) must be distributed about their mean value (which by symmetry is zero) according to Laplace and Gauss' *law of probability*.[1]

M. Chaudesaigues, working in my laboratory, has made the necessary calculations from a series of positions observed in one of my gamboge preparations ($a = \cdot 212\,\mu$). The number n of displacements having projections lying between two successive multiples of $1\cdot 7\,\mu$ (corresponding to 5 millimetres on the squared paper used) are indicated in the following table:—

Projections (in μ) lying between :—	First series.		Second series.	
	n Found.	n Calculated.	n Found.	n Calculated.
0 and 1·7	38	48	48	44
1·7 ,, 3·4	44	43	38	40
3·4 ,, 5·1	33	40	36	35
5·1 ,, 6·8	33	30	29	28
6·8 ,, 8·5	35	23	16	21
8·5 ,, 10·2	11	16	15	15
10·2 ,, 11·9	14	11	8	10
11·9 ,, 13·6	6	6	7	5
13·6 ,, 15·3	5	4	4	4
15·3 ,, 17·0	2	2	4	2

Another and still more striking verification, which was suggested to me by Langevin, is obtained by shifting the

[1] That is to say, out of $\mathfrak{N}$ segments considered,

$$\mathfrak{N} \int_{x_1}^{x_2} \frac{1}{\sqrt{2\pi}} \cdot \frac{1}{X} \, e^{-\frac{x^2}{X^2}} \, dx,$$

will have a projection lying between x_1 and x_2 (the mean square X^2 being measured as above).

observed horizontal displacements in directions parallel to themselves, so as to give them all a common origin.[1] The extremities of the vectors obtained in this way should distribute themselves about that origin as the shots fired at a target distribute themselves about the bull's-eye. This is seen in the figure given below (Fig. 10), on which 500 of my observations with grains of radius $\cdot 367 \, \mu$ are recorded; positions of grains were noted every 30 seconds. The mean

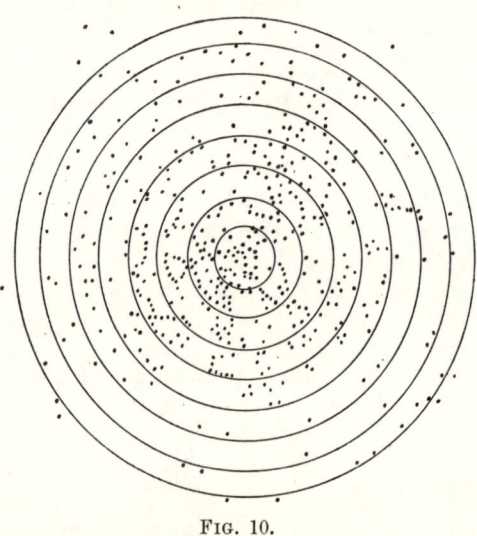

Fig. 10.

square e^2 of these displacements was equal to the square of $7 \cdot 84 \, \mu$. The circles marked in the figure have radii

$$\frac{e}{4}, \; \frac{2e}{4}, \; \frac{3e}{4}, \; \ldots \text{etc.}$$

Here again we have a quantitative check upon the theory; the laws of chance enable us to calculate how many points should occur in each successive ring. In the table on the following page, alongside the probability P that the end point of a displacement should fall in each of the rings, are given the numbers n calculated and found for 500 displacements observed.

[1] This comes to the same thing as considering only grains starting from the same point.

LAWS OF THE BROWNIAN MOVEMENT

Displacement between:—	P for each ring.	n Calculated.	n Found.
0 and $\frac{e}{4}$	·063	32	34
$\frac{e}{4}$,, $2\frac{e}{4}$	·167	83	78
$2\frac{e}{4}$,, $3\frac{e}{4}$	·214	107	106
$3\frac{e}{4}$,, $4\frac{e}{4}$	·210	105	103
$4\frac{e}{4}$,, $5\frac{e}{4}$	·150	75	75
$5\frac{e}{4}$,, $6\frac{e}{4}$	·100	50	49
$6\frac{e}{4}$,, $7\frac{e}{4}$	·054	27	30
$7\frac{e}{4}$,, $8\frac{e}{4}$	·028	14	17
$8\frac{e}{4}$,, $9\frac{e}{4}$	·014	7	9

A third verification is to be found in the agreement established between the values calculated and found for the quotient $\frac{d}{e}$ of the mean horizontal displacement d by the mean quadratic displacement e. By a line of reasoning quite analogous to that (p. 55) which gives the mean speed G in terms of the mean square U^2 of the molecular speed, it is shown that d is very nearly equal to $\frac{8}{9}e$. As a matter of fact, for 360 displacements of grains of radius ·53 μ, I found $\frac{d}{e}$ equal to ·886 instead of ·894 required by the theory.

Further verifications of the same kind might still be quoted, but to do so would serve no useful purpose. In short, the irregular nature of the movement is quantitatively rigorous. Incidentally we have in this one of the most striking applications of the laws of chance.

74.—EARLY VERIFICATIONS OF EINSTEIN'S THEORY (FOR DISPLACEMENTS).—When his formulæ were first published Einstein pointed out that the order of magnitude of the Brownian movement apparently fitted in completely with the requirements of the kinetic theory. Smoluchowski, from his point of view, came to the same conclusion after a searching analysis of the data then available (the fact that the phenomenon is independent of the nature and density of the grains, qualitative observations on the increase in the agitation as the temperature rises or the radius becomes smaller, rough measurements of displacement for grains of the order of 1 micron).

From this it was undoubtedly possible to conclude that the Brownian movement is certainly not more than 5 times more active and certainly not more than 5 times less active than the degree of agitation predicted by theory. This approximate agreement in order of magnitude and qualitative properties immediately gave considerable support to the kinetic theory of the phenomenon, as was clearly brought out by the authors of that theory.

Until 1908 we do not find in the published literature any verification or attempt at verification that adds anything to the information embodied in the conclusions of Einstein and Smoluchowski.[1] About this time a very interesting though partial verification was attempted by Seddig.[2] This author compared, at various temperatures, the displacements undergone in successive tenths of a second by ultra-microscopic grains of cinnabar, which were supposed to be very nearly equal in size. If Einstein's formula is correct, the

---

[1] I cannot even except the work published by Svedberg on the Brownian movement [Zeit. für Electrochemie, *t.* XII., 1906, pp. 853 and 909; Nova Acta Soc. Sc., Upsala, 1907] for the following reasons :—

(*i.*) The lengths given as displacements are 6 or 7 times too great, which, even supposing they were correctly defined, would not contribute any particular advance, particularly to Smoluchowski's discussion of the subject.

(*ii.*) Svedberg believed, which is a much more serious matter, that the Brownian movement becomes *oscillatory* for ultramicroscopic grains. He measured the wave length (?) of this motion and compared it with Einstein's displacement. It is obviously impossible to verify a theory on the basis of a phenomenon which, supposing it to be correctly described, *would be in contradiction to that theory.* I would add that the Brownian movement does not show an oscillatory character on any dimensional scale.

[2] Physik. Zeitschr. Vol. IX., 1908, p. 465.

mean displacements d and d' at the temperature T and T' (viscosities ξ and ξ') should be to one another in the ratio :—

$$\frac{d'}{d} = \sqrt{\frac{T'}{T}}\sqrt{\frac{\xi}{\xi'}}$$

or, for the temperature interval 17°—90° C.,

$$\frac{d'}{d} = \sqrt{\frac{273+90}{273+17}} \cdot \sqrt{\frac{\cdot 011}{\cdot 0032}} = 1\cdot 12 \times 1\cdot 86 = 2\cdot 05.$$

Experiment gives 2·2. The discrepancy is well within the possible error.

Seddig's approximate measurements bring out the influence of viscosity much more than that of the temperature (the effect of the latter in the example quoted is 7 times smaller than the viscosity influence, and it would be difficult to make it very apparent).[1]

Having in my possession some grains of accurately known radius, I was able, at about the same period, to undertake *absolute*[2] measurements and to inquire whether the quotient $\frac{t}{X^2} \cdot \frac{RT}{3\pi a \xi}$, which should be equal to Avogadro's number N according to Einstein's equation, has actually a value independent of the nature of the emulsion and sensibly equal to the value found for N.

That such is actually the case appeared at the time to be far from certain.[3] An attempt by V. Henri to settle the question by a kinematographic experiment, in which for the first time precision was possible,[4] had just led to results distinctly unfavourable to Einstein's theory. I draw attention to this fact because I have been very much struck by the readiness with which at that time it was assumed that the theory rested upon some unsupported hypothesis. I am convinced by this of how limited at bottom is our faith in

---

[1] It is often stated that the Brownian movement may be seen to become more active as the temperature is raised. Actually, from mere inspection, we could affirm nothing if the viscosity did not diminish.

[2] Comptes Rendus, Vol. CXLVII., 1908; Ann. de Ch. et Phys., Sept., 1909, etc.

[3] Compare, for instance Cotton, Revue de Mois (1908).

[4] Comptes Rendus, 1908, p. 146. The method was quite correct and had the merit of being then used for the first time. I do not know what source of error falsified the results.

theories ; we regard them as instruments useful in discovery rather than actual demonstrations of fact.

As a matter of fact, after the completion of the first series of measurements of displacements it became clear that Einstein's formula is accurate.

75.—CALCULATION OF THE MOLECULAR MAGNITUDES FROM THE BROWNIAN MOVEMENT.—I have carried out personally, or directed in others, several series of measurements, varying the experimental conditions as much as I was able, particularly the viscosity and the size of the grains. The grains were picked out in the camera obscura,[1] the microscope being vertical, which gives the horizontal displacements (measured in a micrometer objective). The positions of the grains were generally marked off at 30-second intervals, four positions being obtained for each grain.

I have worked out the method with the help of M. Chaudesaigues, who wished to undertake (series II. and III.) measurements with the grains ($a = ·212\,\mu$), which had given me a good value for N from their vertical distribution. He used a dry objective (Cotton and Moulton's ultra-microscopic arrangement). The following series were obtained with an immersion objective, which permits of a better control of the temperature of the emulsion (temperature variations are important because of the viscosity changes they cause). I obtained the values in series IV. (mastic) in collaboration with M. Dabrowski ; series VI. (in which the liquid was very viscous, X being of the order of 2 μ in five minutes) in collaboration with M. Bjerrum. Series V. refers to two very large mastic grains (obtained in a manner to be described later) ; their diameters were measured directly in the camera lucida and they were suspended in a urea solution of the same density as mastic.

The following table, in which is given, for each series, the mean value of the viscosity ξ, the radius a of the grains, their mass m, and the approximate number n of the displacements recorded, summarises the experiments described above :—

---

[1] It is a matter of real difficulty not to lose sight of the grain as it incessantly rises and sinks. *Vertical* displacements were measured in series VI. only.

LAWS OF THE BROWNIAN MOVEMENT

100 ξ	Nature of the Emulsion.	Radius of the Grains.	Mass $m \times 10^{15}$.	Displacements Recorded.	$\dfrac{N}{10^{22}}$.
1	I. Gamboge grains . .	μ ·50	600	100	80
1	II. Gamboge grains . .	·212	48	900	69·5
4 to 5	III. The same grains in sugar solution (35 per cent.) (temperature only roughly known) . .	·212	48	400	55
1	IV. Mastic grains . .	·52	650	1,000	72·5
1·2	V. Very large grains (mastic) in urea solution (27 per cent.)	5·50	750,000	100	78
125	VI. Gamboge grains in glycerine ($\frac{1}{10}$ water) . .	·385	290	100	64
1	VII. Gamboge grains of very uniform equality . . (Two series) . .	·367 —	246 —	1,500 120	68·8 64

It may be seen from the table that the extreme values of the masses bear a ratio to one another of more than 15,000 to 1, and that the extreme values of the viscosities are in the ratio of 1 to 125. Nevertheless, whatever the nature of the intergranular liquid or of the grains, the quotient $\dfrac{N}{10^{22}}$ remains in the neighbourhood of 70, as in the vertical distribution experiments.[1] This remarkable agreement proves the rigorous accuracy of Einstein's formula and in a striking manner confirms the molecular theory.

The most accurate measurements (series VII.) refer to the most equal set of grains that I have prepared. The prepara-

[1] To these results might be added Zangger's measurements [Zurich, 1911], which were published later. They were obtained from measurements of the *lateral* displacements of mercury droplets sinking through water. The measurements are of interest in that they could be made to refer to a single drop, the radius of which could be obtained from its rate of fall. But this application of Stokes' law to a liquid sphere falling through a liquid is not permissible without a correction that affects the result found for $\dfrac{N}{10^{22}}$ (60 to 79), and which, according to a calculation by Rybczinski, increases that result by about 10 units.

tion and the objective (immersion) were surrounded by water, thus enabling temperature (and consequently viscosity) to be measured accurately. The illuminating beam, of sufficiently feeble intensity, was filtered through a trough of water. The emulsion was very dilute. The microscope was focussed upon the level (6 μ above the bottom) at the height h such that a grain of the size under consideration had the same probability of being above or below it. In order not to be tempted to choose grains which happened to be slightly more visible than the rest (those, that is to say, which were slightly above the average size), which would raise the value of N a little, I followed the first grain that showed itself in the centre of the field of vision. I then displaced the preparation laterally by 100 μ, once more followed the first grain that showed in the centre of the field at the height h, and so on. In this way I obtained the value 69. A source of error has, however, been pointed out to me by M. Constantin.[1] This young physicist noticed, during the course of some measurements on some preparations only a few microns thick, that the proximity of a boundary checked the Brownian movement. (Einstein's theory presupposes an unlimited fluid.) Working at a sufficient distance from the walls with the grains that I had used, he obtained the value $N = 64 \times 10^{22}$; unfortunately the number of observations (about 100) was too small. These measurements will be repeated.

76.—MEASUREMENTS OF THE ROTATIONAL BROWNIAN MOVEMENT (LARGE SPHERULES).—We have seen that Einstein's generalised theory is applicable to the rotational Brownian movement, in which case the formula becomes

$$\frac{A^2}{t} = \frac{R \cdot T}{N} \cdot \frac{1}{4\pi a^3 \zeta},$$

where A^2 stands for the third of the mean square of the angle of rotation in time t.

---

[1] René Constantin, killed in action, 1916, author of a remarkable thesis (C. R. 1914). I have already mentioned that he had verified the extension of Van der Waal's law to concentrated emulsions, finding a negative value for the constant a. He also suggested the application of Smoluchowski's theory of fluctuations (para. 81) to such emulsions by measuring the fluctuations of the number of grains in a small known volume. Preliminary measurements give $N = 60 \times 10^{22}$.

In verifying this formula we check at the same time the estimates of probability that figure in its demonstration and which we meet with whenever we require to establish *equipartition of energy* ; in this particular case this means equality between the mean energies of rotation and of translation. The same difficulties that were met with above (41 to 43), with regard to the limits of applicability of such equipartition, increase the desirability of a verification.

The formula, however, indicates a mean rotation of about 8 degrees in the one-hundredth of a second, for spheres $1\,\mu$ in diameter ; such rotation is too rapid to be perceived (more especially as no distinguishing marks can be noticed on such small spherules), much less to be measured. And as a matter of fact, this rotation has never been made the subject of any experimental study, even qualitative.

I have overcome the difficulties in the way by preparing very large gamboge and mastic spherules. This was done by precipitating the resins from alcoholic solution, not in the usual way by the sudden addition of a large excess of water (which produces grains of diameter generally less than 1 micron), but by causing the precipitating water to penetrate slowly and progressively into the resin solution. This was managed by very slowly running pure water from a funnel with a very slender spout under an alcoholic solution of resin (dilute), which is steadily forced up by it. A zone is established between the two liquids across which they diffuse into each other, and the grains that are formed in the zone have diameters of quite a dozen microns. They therefore soon become so heavy that they sink, in spite of their Brownian movement, passing downwards through the pure water, where they are washed, to the bottom of the apparatus, from which they can be recovered after decantation of the supernatent liquid. In this way I have precipitated all the resin in alcoholic gamboge and mastic solutions in the form of spheres having diameters as high as $50\,\mu$. These large spheres look like glass balls, yellow with gamboge, colourless with mastic, which are readily broken up into irregular fragments. They often appear to be perfect and, like lenses, they produce real, recognisable images of the source of light which illuminates the preparation (an Auer mantle, for instance). Frequently, however, *they contain inclusions* [1] *by means of which the rotational Brownian movement may easily be perceived.*

Unfortunately the weight of these grains keeps them always very near the bottom of the vessel, where their Brownian movement may possibly be affected by cohesion phenomena. I have therefore tried, using solutions of various suitable substances, to render the intergranular liquid of the same density as the grains themselves. With nearly all the substances, however, a complication arose, in that the concentration necessary to keep the grains just suspended without rise or fall was sufficient to cause the grains to coagulate into grape-like clusters. This provides a very pretty illustration of the phenomenon of coagulation and its mechanism, which is not very easily demonstrated with ordinary colloidal solutions (in which the grains are ultra-microscopic). With urea alone does coagulation not take place.

I have thus been able to follow the agitation of the grains in water containing 27 per cent. urea (series V. in the preceding table). At the same time it has been found possible to measure, more or less roughly, their rotation. In doing this I marked, at equal intervals of time, the successive positions of particular granular inclusions. This enabled me subsequently to fix the orientation of the spheres at each instant and to calculate approximately their rotation from one instant to another. Calculations based on about 200 measurements with spheres 13 μ in diameter gave me, by the application of Einstein's formula, the value 65×10^{22} for N, the probable exact value being 69×10^{22}. In other words, starting from the latter value for N, we should expect to find for $\sqrt{\overline{A^2}}$, in degrees per minute, the value $14°$; by experiment we find $14·5°$.

[1] These inclusions do not appreciably affect the density of the grains; in an aqueous urea solution mastic grains remain in suspension, in solutions containing equal quantities of urea, whether they do or do not contain inclusions. I have also investigated the nature of the inclusions, which probably consist of a viscous paste containing traces of alcohol.

In exceptional cases a grain is sometimes found to be made up of two spheres united about a small circle, an effect clearly due to the fusion of two spheres whilst they are still growing from their respective nuclei. The dual question of the initial formation of the nuclei and their rate of growth has an interest outside the scope of the present inquiry.

The discrepancy is well below the possible error introduced by the somewhat loose approximations used in connection with the measurements and in making the calculations. The agreement is still more striking because *a priori* we know nothing even of the order of magnitude of the phenomenon. The masses of the grains observed were 70,000 times greater than those of the smallest studied in the determination of vertical d'stribution.

77.—THE DIFFUSION OF LARGE MOLECULES.—To carry out our intention of establishing the various laws deduced by Einstein on an experimental basis, it only remains to study the diffusion of emulsions and to see whether the value of N derived from the equation

$$N = \frac{RT}{D} \cdot \frac{1}{6\pi a \xi}$$

agrees with that already found.

In this connection it is proper to refer to the application by Einstein himself of his formula to the diffusion of sugar into water. In applying the formula to this particular case it is assumed—(i.) that sugar molecules may be regarded as very nearly spherical, and (ii.) that Stokes' law is applicable to them. (It is therefore not surprising that the value expected was not obtained.)

Making these assumptions, the equation in question, applied to the case of sugar at 18° C., becomes [1]

$$aN = 3\cdot 2 \times 10^{16}$$

We do not, however, know what radius may be assigned to the sugar molecule, for we cannot calculate it by the process available for volatile substances.

It may be pointed out, as has been done above (para. 47), that we obtain some indication of the "true" volume ($\frac{4}{3}\pi a^3 N$) of the molecules making up a gramme molecule

[1] For we know (para. 37: note) that R is equal to $83\cdot 2 \times 10^6$, that D is equal to $\frac{\cdot 33}{86,400}$, (para. 70 : note), and that T is equal to $(273 + 18)$ °C. Moreover, the viscosity at this temperature of the *pure* intermolecular water, to which the reasoning applies (and not the total viscosity of the sugar solution), is $\cdot 0105$ (para. 48 : note).

of sugar by measuring the volume (208 c.c.) occupied by that quantity of sugar in the crystalline state. Einstein has very neatly overcome the difficulties in the way by calculating this volume from the viscosity of the sugar solution. He did this by showing, from the laws of hydrodynamics, that an emulsion of spherules should be more viscous than the pure intergranular liquid and that the relative increase in viscosity $\frac{\xi'-\xi}{\xi}$ is proportional to the quotient $\frac{v}{V}$, of the volume V of the emulsion into the true volume v of the spherules present therein. The first calculations actually indicated pure and simple equality between $\frac{\xi'-\xi}{\xi}$ and $\frac{v}{V}$.

Extrapolating this theory, once established for emulsions, to the case of a sugar solution, Einstein obtained in an approximate manner the true volume of the molecules making up a gramme molecule of sugar. Using the value already obtained for the product aN, he found (1905) the value 40×10^{22} for the number N.[1]

A few years later M. Bancelin, working in my laboratory, set himself to verify the formula given for the relative increase in viscosity (which promised to be easy with gamboge or mastic emulsions). It was at once apparent that the increase predicted by the formula was too small.

On hearing of this lack of agreement Einstein noticed that

---

[1] The results of a subsequent verification of the diffusion formula by Svedberg [Zeit. für Phys. Chem., Vol. LXII., 1909, p. 105] may be compared with this. Svedberg used colloidal gold solutions, the grains being invisible under the microscope. The diameter of the grains, calculated, according to Zsygmondy's method, to be $\cdot 5 \times 10^{-7}$, and the co-efficient of diffusion (equal to $\frac{4}{5}$ that of the sugar solutions), should give about 66×10^{22} for N. The high degree of uncertainty involved in the measurement (and even in the definition) of the radii of *invisible* granules (which are probably sponge-like bodies of *widely differing* bulk) renders these results on the whole less convincing than those deduced by Einstein from the diffusion of molecules that were not invisible, *very much less massive,* and identical among themselves.

Svedberg has also carried out certain *relative* measurements, wherein he compares the diffusions of two colloidal gold solutions, the grains in the one being (on the average) 10 times smaller than the grains in the other; from colorimetric measurements he drew the conclusion that 10 times as many more small grains than large pass through identical membranes in the same time. This is just what would be expected from the formula (supposing always that the pores in the parchment were sufficiently large).

an error had occurred, not in the reasoning, but in the calculation, and that the correct formula should be

$$\frac{\xi' - \xi}{\xi} = 2{\cdot}5\,\frac{v}{V},$$

which agrees with the measurements. The corresponding value for N is now found to be

$$65 \times 10^{22},$$

which agrees remarkably well with the accepted value. This forces us to take the view that sugar molecules possess a more or less compact structure, even if they are not spherical, and that Stokes' law is, moreover, applicable to molecules which are certainly relatively large, although their diameters do not exceed the thousandth of a micron.

78.—FINAL EXPERIMENTAL PROOF : THE DIFFUSION OF VISIBLE GRANULES.—As he himself demonstrated, Einstein's diffusion equation

$$D = \frac{RT}{N} \cdot \frac{1}{6\pi a \xi},$$

which can be only approximate for molecules, happens to be rigorously obeyed by emulsions. In fact, since this equation is the necessary consequence of Stokes' law and the vertical distribution law, it may be regarded as verified in the domain in which I have shown that these laws apply.

Direct measurements of diffusion, however, if carried out in such a way as to extend that domain, have a certain interest.

When, therefore, M. Léon Brillouin made known to me his wish to complete the experimental verification of Einstein's theory by studying the diffusion of emulsions, I suggested to him the following method, which makes use of the obstacle that prevented my studying permanent equilibrium in pure glycerine, in which the grains stick to the glass walls of the containing vessel when they chance to come in contact with it (para. 66 : note).

Consider a vertical glass partition enclosing an emulsion, initially of uniform distribution, composed of gamboge grains in glycerine, the number of grains per unit volume being n. The partition, which behaves as though it were a perfect "absorber," captures all grains brought by chance Brownian movements into contact with it, so that the emulsion becomes steadily weaker by diffusion towards the glass, while the number $\mathfrak{N}$ of the grains collected by unit surface steadily increases. The variation of $\mathfrak{N}$ with the time will determine the coefficient of diffusion.

The absorbing partition observed will be the lower surface of the object-glass confining a preparation maintained vertically at an absolutely constant temperature. The thickness of the preparation will be sufficiently great to ensure that during observations extending over several days the absorption by the cover-glass will be throughout what it would be if the emulsion extended to infinity.[1]

The following approximate line of reasoning enables us to deduce the coefficient of diffusion D from the measurements taken.

Let X^2 throughout be the mean square (equal to $2Dt$) of the displacement during time t that elapses from the beginning of an experiment. No great error will be introduced if we assume that each grain has undergone, either towards the absorbing partition or in the opposite direction, the displacement X. The number $\mathfrak{N}$ of the grains stopped by unit surface during the time t is then clearly

$$\mathfrak{N} = \frac{1}{2} nX,$$

from which we get, replacing X by $\sqrt{2Dt}$,

$$\frac{\mathfrak{N}^2}{t} = D \frac{n^2}{2},$$

or

$$D = \frac{2}{n^2} \cdot \frac{\mathfrak{N}^2}{t},$$

which is the required coefficient of diffusion.

[1] The grains, being slightly less dense than glycerine, slowly rise (about 1 millimetre in two weeks at the temperature of experiment). This fact has no influence on $\mathfrak{N}$ if the preparation is deep enough to ensure that the surface studied always remains above the lower layers that are impoverished by this rising of the grains.

LAWS OF THE BROWNIAN MOVEMENT

M: Léon Brillouin carried out the experimental work and obtained measurements—a work of considerable difficulty—with much skill. Gamboge grains, equal in size (radius ·52 μ), freed by desiccation from intergranular water, were treated for a long time with glycerine, a dilute uniformly distributed emulsion containing $7·9 \times 10^8$ grains per cubic centimetre being obtained (the volume of the grains thus did not come within $\frac{2}{1,000}$ of that of the emulsion). Diffusion took place in a thermostat constant at $38·7°$ C., at which temperature the viscosity of the glycerine employed was 165 times that of water at $25°$ C. Twice a day the same portion of the partition to which the grains were adhering was photographed and the grains counted on the negatives. The diffusion was followed in six preparations, each during the course of several days.[1]

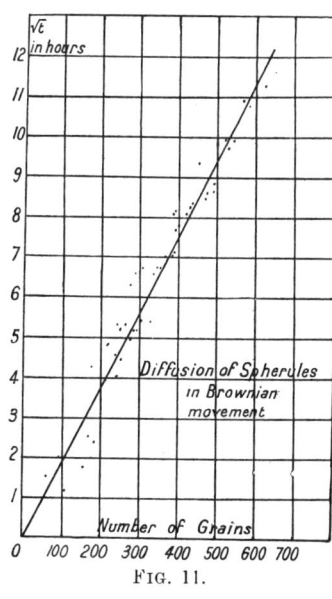

Fig. 11.

Examination of the series of negatives showed that the square of the number of grains fixed is roughly proportional to the time, so that, plotting the results so that the abscissæ represent the values of $\mathfrak{P}$ and the ordinates the time $\sqrt{t}$, the points representing the measurements fall roughly on a straight line passing through the origin, as is shown in the adjoining figure. The coefficient D, equal to $\frac{2}{n^2}\frac{\mathfrak{P}^2}{t}$, follows

[1] M. Brillouin has examined qualitatively preparations kept at the melting point of ice, at which temperature the viscosity of glycerine becomes more than 3,000 times that of water. The Brownian movement, which is quite difficult to perceive with the viscosity at its initial value, now appears to be completely arrested. It occurs, nevertheless, and successive photographs show that grains diffuse slowly towards the partition, the number of grains which happen to adhere to it increasing with time in the right way, although it was not possible to wait long enough for accurate measurements to be taken.

at once. It is found to be equal to $2 \cdot 3 \times 10^{-11}$ for the grains employed, deduced from the fixation of several thousand grains; this corresponds with a rate of diffusion 140,000 times slower than that of sugar in water at 20° C.

To verify Einstein's diffusion equation, it only remains to see whether the number $\dfrac{RT}{D} \cdot \dfrac{1}{6\pi a \xi}$ is near 70×10^{22}. As a matter of fact, it is equal to 69×10^{22} to within ± 3 per cent.

79.—SUMMARY.—The laws of perfect gases are thus applicable in all their details to emulsions. This fact provides us with a solid experimental foundation upon which to base the molecular theories. The field wherein verification has been achieved will certainly appear sufficiently wide when we remember:

That the nature of the grains has been varied (gamboge, mastic);

That the nature of the intergranular liquid has been varied (pure water, water containing 25 per cent. urea or 33 per cent. sugar; glycerine, containing 12 per cent. water, pure glycerine);

That the temperature varied (from $-9°$ C. to $+58°$ C.);

That the apparent density of the grains varied (between $-\cdot 03$ and $+\cdot 03$);

That the viscosity of the intergranular liquid varied (in the ratio of 1 to 330);

That the mass of the grains varied (in the enormous ratio of 1 to 70,000) as well as their volume (in the ratio of 1 to 90,000).

From the study of emulsions the following values have been obtained for $\dfrac{N}{10^{22}}$:—

      68·2 deduced from the vertical distribution of grains.
      64   deduced from their translatory displacements.
      65   deduced from observations on their rotation.
      69   deduced from diffusion measurements.

If we wish we may express our results by stating that the mass of the hydrogen atom, *in terms of trillionths of trillionths*

LAWS OF THE BROWNIAN MOVEMENT

of a gramme, has the values 1·47, 1·56, 1·54, and 1·45 respectively.

As we shall see later, other facts imply a discontinuous structure for matter, and, like the Brownian movement, enable us to estimate the masses of the structural units.

CHAPTER V

FLUCTUATIONS

SMOLUCHOWSKI'S THEORY.

THE molecular agitation of which the Brownian movement is the direct manifestation can be inferred from other sets of phenomena that include a constant succession of variable inequalities in microscopic portions of matter in equilibrium.

80.—DENSITY FLUCTUATIONS.—We have already indicated one of these phenomena in speaking of the definite though very feeble thermal inequalities which are produced spontaneously and continuously in spaces of the order of a micron, and which are, indeed, a second aspect of the Brownian movement itself. These thermal fluctuations, of the order of a *thousandth* of a degree for such volumes,[1] seem in practice to be inaccessible to our measurements.

The density of a fluid in equilibrium, like its temperature or molecular agitation, should vary from point to point. A cubic micron, for example, will contain sometimes a larger and sometimes a smaller number of molecules. Smoluchowski has drawn attention to these spontaneous inequalities, and has been able to calculate the fluctuation in density, $\frac{n - n_o}{n}$, n being the chance number of molecules in a volume v of fluid which in the case of rigorously constant and uniform concentration would contain n_o molecules.

To begin with he showed, by the calculus of probability, that the absolute mean value of this fluctuation for a gas or a dilute solution shou'd be equal to $\sqrt{\frac{2}{\pi} \cdot \frac{1}{n}}$. If the density of the gas is the so-called normal density, we see that the mean

[1] According to a calculation by Einstein, based, like the formulæ that have already been verified, on the kinetic theory of emulsions.

variation, for volumes of the order of a cubic centimetre, is of the one thousand-millionth order only. It becomes of the order of one-thousandth for the smallest cubes resolvable by the microscope. Whatever the density of the gas, the variation will be about 1 in 100 if the volume considered contains 6,000 molecules and 10 in 100 if it contains 60.[1] It should thus be readily measured and the theory verified with dilute emulsions.[2]

81.—CRITICAL OPALESCENCE.—No longer confining himself to the case of rarefied substances, Smoluchowski succeeded a little later, in a most remarkable memoir,[3] in calculating the mean density fluctuation for any fluid whatever, and proved that, even with condensed fluids, the fluctuations should become noticeable in spaces visible under the microscope when the fluid is near the critical state.[4] He thus succeeded in explaining the enigmatic *opalescence*[5] which is always shown by fluids in the neighbourhood of the critical state.

This opalescence, which is absolutely stable, indicates a permanent condition of fine grained heterogeneity in the fluid. Smoluchowski explains it as being due to the magnitude of the compressibility (infinite at the critical point

---

[1] I had hoped to perceive these fluctuations in dilute solutions of fluorescent substances. I have found, however, that such bodies are destroyed by the light which makes them fluoresce: fluorescence indicates a chemical action (Ann. de Physique, 1918).

[2] Svedberg has in fact been able to conclude that the formula is verified in dilute emulsions of gamboge (*Zeit. Phys. Chem.*, 1910).

[3] Acad. des Sc. de Cracovie, December, 1907.

[4] It is known that for every fluid there is a temperature above which it is impossible to liquefy it by compression; that temperature is the *critical temperature* (31° C. for carbon dioxide). Similarly, there is a pressure above which a gas cannot be liquefied by cold; that pressure is the *critical pressure* (71 atmospheres for carbon dioxide). A fluid is in the critical state when it arrives at its critical temperature under its critical pressure. At the point representing the critical state in a p, v, T diagram the isothermal shows a point of inflexion, the tangent at that point being parallel to the volume axis (at this point $\frac{dp}{dv}$ is nothing and the compressibility is infinite).

[5] A liquid is opalescent if the path of a beam of light is visible in it, as in soapy water or air charged with smoke. The light thus seen is distinguished from *fluorescent* light in that, when analysed in the spectroscope, it contains no colours that are not found in the illuminating beam, although its tint is generally more bluish owing to change in the distribution of intensities (it is also distinguishable by the fact that, being completely polarised, it fails to reach an eye observing it at right angles to the pencil through a suitably orientated analyser).

itself) which enables contiguous regions of notably different density to be nevertheless almost in equilibrium with each other. Hence, owing to the molecular agitation, the formation of dense swarms of molecules, diffuse in contour, will be facilitated. These swarms will break up but slowly, while at the same time others will be forming elsewhere and will produce opalescence by causing lateral deviation of the light.

In more precise terms, the theory shows first that, when far from the critical point, the mean square of the fluctuation in volume ϕ is very nearly

$$-\frac{RT}{N} \cdot \frac{1}{\phi v_0} \cdot \frac{dv}{dp},$$

$\frac{\partial v}{\partial p}$ being the (isothermal) compressibility and v_0 the specific volume of the fluid. But near the critical point the formula ceases to be valid and the theory gives, as the absolute mean value of the fluctuation in the volume, which under uniform distribution would contain n_0 molecules, very nearly the inverse of the fourth root of n_0; this gives 2 per cent. in a cube containing 10^8 molecules. For most fluids in the critical state the side of such a cube is of the order of 1 micron. The heterogeneity is then much more accentuated than in a gas, and we can understand why opalescence, which is in reality always more or less in existence, then becomes very marked.

82.—EXPERIMENTAL VERIFICATION OF THE THEORY OF FLUCTUATIONS.—In the first place, having once obtained the osmotic compressibility of a concentrated emulsion, which is, as it were, a fluid with visible molecules,[1] one should be able to see whether the fluctuations at a given level obey Smoluchowski's formula; this contains N, and we would then have a new method for finding its value. This method was suggested by R. Constantin, and preliminary measurements gave him the value $N = 60 \times 10^{22}$.

With regard to the critical opalescence, the theory must be completed befo e a calculation can be made from diffused light (Keesom).

[1] J. Perrin (C. R., 158, 1914); R. Constantin (*ibid.*, p. 16).

FLUCTUATIONS 137

A theory due to Rayleigh gives the quantity of light diffused in a given beam of light by a small transparent particle placed in a medium of different refractive index. The diffusion is greater the more refrangible the incident light. With white light the diffused light should therefore be bluish, which is, of course, the case.

More accurately, as long as the dimensions of the illuminated particle may be regarded as small compared with the wave length of the incident light, the intensity of the diffused light is inversely proportional to the fourth power of that wave length, but directly proportional to the square of the volume of the particle and to the square of the relative difference in refractive index.[1]

If, as actually happens in the case of density fluctuations, the particle which deviates the light is composed of the same substance as the surrounding medium, this relative variation in index is proportional to the relative variation in density,[2] that is, to the fluctuation $\frac{n - n_o}{n_o}$, the mean quadratic value of which has been given by Smoluchowski. Summing all the intensities thus separately due to the small sections composing a perceptible volume of fluid, we find that the intensity i of the light diffused by a cubic centimetre at right angles to the incident rays is

$$i = \frac{\pi^2}{18} \cdot \frac{1}{\lambda^4} \cdot \frac{RT}{N} \cdot (\mu_o^2 - 1)^2 (\mu_o^2 + 2)^2 \cdot \frac{1}{-v_o \cdot \frac{\partial p}{\partial v_o}},$$

where μ_o is the refractive index (mean) of the fluid for the light used of wave length λ (in "free space" or *vacuo*),

[1] At right angles to the incident light, this intensity is given by the expression

$$2\pi^2 \cdot \frac{\phi^2}{\lambda^4_o} \cdot \left(\frac{\mu - \mu_o}{\mu_o}\right)^2,$$

ϕ being the volume, λ_o the wave length in the medium outside the particle and μ_o and μ the refractive indices in that medium and in the particle.

[2] This follows from the law of refraction (Lorentz), according to which $\frac{1}{d} \cdot \frac{\mu^2 - 1}{\mu^2 + 2}$ is constant for any fluid.

v_o the specific volume of the fluid, and $\frac{\partial p}{\partial v^o}$ its compressibility (isothermal).

All the quantities in the above equation are measurable except N; a comparison of the value of N derived thus with the value obtained already will therefore enable us to check the theories of Smoluchowski and Keesom.

An examination of the fine series of measurements recently carried out on ethylene by Kamerlingh Onnes and Keesom will be found to provide the required test. The critical temperature (absolute) was $273 + 11·18°$; the opalescent light was quite blue even at $11·93°$. At this temperature the ratio of the intensities of opalescence for incident light of the same intensity in the blue and yellow (lines F and D) was 1·9, but little different from the ratio 2·13 of the fourth powers of the vibration frequencies of the two colours.

At the same temperature measurements in yellow light gave, per centimetre cube illuminated and for incident light of intensity 1, an intensity of opalescence varying between ·0007 and ·0008. The compressibility is known from Verschaffelt's measurements. Keesom's formula then gives, for Avogadro's number N, a value in the neighbourhood of 75×10^{22} with a possible error of 15 per cent., which is in very good agreement with the probable value.

Analogous considerations can be applied to the opalescence always shown by liquid mixtures (water and phenol, for example) in the neighbourhood of the point of critical miscibility.[1] Opalescence in this case indicates a permanent condition of fluctuation in composition from one point to another in the mixture. The theory of these fluctuations, which is a little more difficult than in the above case, has been given by Einstein (using the conception of work done in

---

[1] At all temperatures below 70° C. the mutual solubilities of water and phenol are limited; two layers of liquid are produced, containing unequal amounts of phenol. As the temperature rises, the difference between the two layers becomes less and less, until at 70° C. the concentration of phenol becomes equal to 36 per cent. throughout; the dividing surface then disappears and the point of critical miscibility is reached. At all higher temperatures miscibility is complete and two layers of different composition can no longer remain in equilibrium in contact with each other.

FLUCTUATIONS 139

separating the constituents instead of the idea of work done in compression). The equation [1] he has obtained, assuming it to be exact, again allows us to find N from measurable quantities, but in this case the determination has not yet been carried out.

83.—THE BLUENESS OF THE SKY.—We have applied the formulæ of Smoluchowski, Keesom, and Einstein in the neighbourhood of the critical point. They are equally applicable to the case of a gaseous substance. We will suppose that the gas is pure, or at least, if it is a mixture, that its components have the same refracting power (which is sensibly the case for air), so that fluctuations in composition will have a negligible influence in comparison with density fluctuations. In this case, making use of Boyle's law, the product $\left(-v_o \dfrac{\partial p}{\partial v_o}\right)$ becomes equal to $\dfrac{1}{p}$; further, the refractive index being very nearly equal to 1, we can replace $(\mu_o^2 + 2)$ by 3, and Keesom's equation becomes

$$i = \frac{\pi^2}{2\lambda^4} \cdot \frac{RT}{N} \cdot \frac{1}{p} \cdot (\mu_o^2 - 1)^2.$$

The quantity of light thus emitted laterally by 1 cubic centimetre of gas is extremely small, because of the feeble refractive power of gases (μ_o^2 is very little greater than 1). Cabanes has succeeded, however, in measuring accurately the faint light diffused laterally by a small element of a pure gas (argon) under intense illumination. This gives a new accurate determination of N, yielding the value 69×10^{22}.

Furthermore, the total light produced by a very large volume may become perceptible, and the blue light that comes from the sky in daytime can thus be explained. In this way we arrive at the conclusion already reached by Lord Rayleigh.[2]

We know that a beam of light has a visible track when traversing a medium charged with dust. To this lateral diffusion is due the visibility of a "sunbeam" in the air. The phenomenon still persists as the dust particles become increasingly smaller (and it is this fact that makes

[1] Ann. der Phys., Vol. XVI., 1910, p. 1572.
[2] Phil. Mag., Vol. XLI., 1871, p. 107, and Vol. XLVII., 1899, p. 375.

ultra-microscopic observation possible), but the diffracted opalescent light turns to blue, light of shorter wave length thus undergoing the greater diffraction. It is, moreover, polarised in the plane passing through the incident ray and the eye of the observer.

Rayleigh supposed that the molecules themselves behave like the dust particles just visible under the microscope and that the origin of the colour of the sky lies in them. In agreement with this hypothesis, it is found that the blue light from the sky, when observed in a direction perpendicular to the sun's rays, is strongly polarised. It is, moreover, difficult to believe that it is a question of actual dust particles, for the blueness of the sky is not diminished in the slightest at the height of 2,000 or 3,000 metres, which is well above most of the dust that contaminates the air near the earth. We may therefore conclude that we have here a means of counting the diffracting molecules which enable us to see a given portion of the sky and in consequence a means for obtaining N.

Rayleigh did not restrict himself to this merely qualitative conception, but calculated, while developing the elastic theory of light, the relation that should, on his hypothesis, obtain between the intensity of the direct solar radiation and that of the light diffused by the sky. Let us suppose that we are observing the sky in a direction the zenith-distance of which is a and which makes an angle β with the solar rays; the illuminations e and E obtained in the field of an objective pointed successively towards this region of the sky and towards the sun should be, for each wave length λ, in the ratio :—

$$\frac{e}{E} = \pi^3 \omega^2 M \cdot \frac{p}{g} \cdot \frac{1 + cos^2\beta}{cosa} \left(\frac{\mu^2 - 1}{d}\right)^2 \cdot \frac{1}{\lambda^4} \cdot \frac{1}{N},$$

where ω represents the apparent semi-diameter of the sun, p and g the atmospheric pressure and the acceleration due to gravity at the point of observation, M the gramme-molecular weight of air (28·8 gramme), $\frac{\mu^2 - 1}{d}$ the refractive power of air (Lorentz), and N Avogadro's constant. Langevin obtained the same equation (with μ^2 replaced by the dielectric constant K) during the course of development of a simple electro-magnetic theory. In each case the preceding formula was obtained by summing the intensities of the light diffracted by the individual molecules (assumed to be distributed in an entirely irregular manner).

Identically the same formula is obtained (for $\beta = 90°$) by applying Keesom's equation, as was shown by Einstein.

It follows that the extreme violet of the spectrum should be 16 times more diffracted than the extreme red (the wave length of which is twice as great), and this is well borne out by the actual colour of the sky (which no other hypothesis has succeeded in explaining).

The above formula takes no account of the light reflected by the earth. The brightness of the sky would be doubled by a perfectly reflecting earth (which would be equivalent to the illumination of the atmosphere by a second sun). The reflecting power of the earth entirely covered with snow or by clouds would be little different from ·7, and the brightness of the sky would be 1·7 times that due to the sun alone.

An experimental verification should be possible at a height sufficient to avoid perturbations due to dust (smoke, small drops of water, etc.). The first indication of such a verification was obtained by Lord Kelvin from the early experiments of Sella, who, at the summit of Monte Rosa, compared the brightness of the sun at a height of 40° and the brightness of the sky at the zenith at the same instant and obtained a ratio equal to 5,000,000. This gives for $N \times 10^{-22}$ (allowing for the absence of precision with regard to wave length) a value between 30 and 150. Roughly, the correct order of magnitude was attained.

Bauer and Moulin [1] have constructed an apparatus for making the necessary spectrophotometric comparison and have made some preliminary measurements on Mont Blanc, with, unfortunately, a not very favourable sky. Their comparisons give (for green light) numbers between 45 and 75 for $N \times 10^{-22}$. More recently, however, the remarkable

[1] Killed in action, 1914.

measurements of Fowle [1] have given $N = 60 \times 10^{22}$. Rayleigh's theory is thus verified beyond doubt and the blue colour of the sky is shown to be one of the phenomena through which the discontinuous structure of matter carries its effects into our usual scale of magnitudes.

With regard, however, to the degree of accuracy obtained by Fowle, it is necessary to correct the theory by taking account of the fact that the molecules of air, being anisotropic, cause incomplete polarisation of the luminosity (Strutt). The correction, which has been applied by Cabanes, gives $N = 65 \times 10^{22}$.

84.—CHEMICAL FLUCTUATIONS.—Up to the present we have not attempted to formulate a kinetic theory of chemical reaction ; without going deeply into the matter at this early stage, a few simple remarks may be pertinent.

We will limit ourselves to the consideration of two particularly important and simple types of reaction, types which, in fact, by addition or repetition make up all classes of chemical reaction. On the one hand we have dissociation or the splitting up of one molecule into simpler molecules or into atoms (I_2 into $2I$; N_2O_4 into $2NO_2$; PCl_5 into $PCl_3 + Cl_2$, etc.), which is expressed in general form by

$$A \longrightarrow A' + A'' ;$$

on the other hand we have the inverse phenomenon or the building up of a molecule, expressed by

$$A \longleftarrow A' + A''.$$

If at a given temperature two inverse transformations exactly counterbalance one another :—

$$A \rightleftarrows A' + A'',$$

so that on our scale of observation the quantities of the components remain constant throughout the system, we say that chemical equilibrium has been reached and that no further change will take place.

[1] Astrophysical Journal, 1914

Actually both reactions are taking place, and at each instant an enormous number of molecules are breaking up at certain points while at others an equivalent amount of A is being re-formed. I have no doubt that in microscopic spaces we should be able to see, at a sufficient magnification, an incessant fluctuation in chemical composition. Chemical no less than physical equilibrium in fluids is merely an illusion that masks a continuous cycle of compensating transformations.

A quantitative theory of this chemical Brownian. movement has not yet been developed. But, though only qualitative, the kinetic conception of equilibrium has rendered great services. It is the real basis of the whole of chemical mechanics that is concerned with velocities of reaction (Law of Mass Action).

85.—FLUCTUATIONS IN MOLECULAR ORIENTATION.—The remarkable phenomenon discovered by Mauguin during the course of his splendid work on liquid crystals falls into the same group of phenomena as the Brownian movement and the fluctuations of density and composition.

It has been known, since Lehmann's famous investigations, that there are some liquids which exhibit when in equilibrium the optical symmetry of uniaxial crystals, so that when a film of one of them is examined under the microscope between a polariser and analyser set at the extinction point illumination is re-established, except where the crystalline orientation of the liquid is parallel to the ray of light traversing it. When, however, the light is very intense, we notice that extinction is not absolute for such orientations and that an incessant scintillation, like the swarming of a luminous ant heap, is visible at all points in the field, producing a feeble light that varies rapidly from place to place and from instant to instant.[1] Mauguin at once connected this phenomenon with the Brownian movement, and, indeed, it seems difficult to explain it except on the supposition that the molecular agitation continuously tends

---

[1] This is well shown by para-azoxyanisol, spread out in a thin film between two accurately plane glass surfaces (the crystalline axes being thus fixed perpendicularly to the plane surfaces) and maintained at temperatures lying between 130° C. and 165° C. (outside these temperature limits change of state occurs).

to strain the molecular axes from their positions of equilibrium. Analogous fluctuations should occur during the magnetisation of ferro-magnetic bodies and undoubtedly the theories of ferro-magnetism (P. Weiss) and of liquid crystals will reduce to a common basis.

CHAPTER VI

LIGHT AND QUANTA

Black Bodies.

86.—ANY CAVITY COMPLETELY ENCLOSED BY A MATERIAL AT A UNIFORM TEMPERATURE IS FULL OF LIGHT IN STATISTICAL EQUILIBRIUM.—When a fluid fills an enclosure molecular agitation, which is the more active the higher the temperature, gradually transmits from point to point all thermal actions and the degree of agitation gives a measure of the temperature once equilibrium is established. But we know that, even in the absence of all intermediary matter, the temperature of the space inside an *isothermal* enclosure (an enclosure in which, that is to say, the temperature is uniform) has a definite physical significance; we know that a thermometer always ends by giving the same indication (it arrives, that is to say, at the same final state) at any point whatever in an opaque enclosure surrounded with boiling water, whether the enclosure contains any fluid whatever or whether it is absolutely *empty*. The effect upon the thermometer in the latter case is produced solely by *radiation* from the various points of the enclosing medium.

This radiation is visible or not according to the temperature of the enclosure (an ice-house, an oven, or an incandescent furnace), but its visibility, which is of importance to us alone, has no claim to be regarded as an essential characteristic of the radiation, which is *light* in the general sense of the word and traverses space at the invariable velocity of 300,000 kilometres per second.

When we say that the enclosure is sealed and that it is opaque, we mean that no thermal influence can be exerted by radiation between two objects, one of which is inside and the other outside the enclosure.[1] This is the reason why a thermometer inside the enclosure reaches and persists in a definite invariable state. This does not mean, however, that no subsequent change takes place in the region wherein the indicating thermometer is placed. That region is constantly receiving radiation emitted by the various parts of the enclosure ; the fixed indication shown by the receiving instrument (thermometer), however, proves that it undergoes no further change in property, but maintains itself in a *stationary* condition.

This stationary state in space traversed continually and in all directions by light really represents a permanent condition of extremely rapid changes. Details of them escape us, in spaces and times on our usual dimensional scale, just as the agitation of the molecules in a fluid in equilibrium cannot be perceived, although the latter phenomenon is of a much higher order of magnitude. In fact, the thermal equilibrium in fluids, which has already been studied at length, and the thermal equilibrium of light are in many respects comparable. I now propose to define our conceptions of the latter equilibrium.

I have pointed out that a thermometer invariably registers the same temperature, at all points inside a closed cavity with walls at a fixed temperature, that it would show in contact with the walls themselves. This remains true whether the enclosure is made of porcelain or of copper, whether it is large or small, prismatic or spherical. More generally, whatever the means of investigation employed, we shall find that absolutely no influence is exerted by the nature of the enclosure, its size or shape, on the stationary condition of the radiation at each point ; this state completely determines the only temperature to be recorded within the enclosure.

It follows from this that all directions passing through a given point are equivalent. No arrangement of lenses or mirrors, in the interior of an incandescent furnace, would

[1] It is obviously possible to concentrate light of external origin by means of lenses upon a thermometer suspended within a cavity in a *transparent* block of ice and to make it indicate any desired temperature.

produce the slightest effect; neither temperature nor colour would be altered in the least and no images would be formed. Expressed differently, the point image of a point on a wall would not be distinguishable by any property whatever from any other point inside the furnace. An eye capable of existing at the temperature of the furnace would not be able to distinguish any particular object or outline and would perceive merely a general uniform illumination.

Another necessary consequence of the existence of a stationary *régime* is that the *density* W of the light (quantity of energy contained in 1 cubic centimetre) will have a definitely fixed value for each temperature. Similarly, if we consider within the enclosure a flat closed contour 1 square centimetre in area, the quantity of light passing across the contour in one second, say from the left towards the right of an observer lying along the edge of the contour and looking towards its interior, is at each instant equal to the quantity of light passing in the same time in the opposite direction and has a perfectly definite value E, which is proportional to the density W of the light in equilibrium at the given temperature. More precisely, if c stands for the velocity of light, it appears, as the result of a simple integration, that E is equal to $\frac{Wc}{4}$. It is clear, moreover, that strictly speaking the quantities of light E or W undergo *fluctuations* (which are negligible on the dimensional scale with which we are concerned).

87.—BLACK BODIES : STEFAN'S LAW.—A knowledge of the density of the light in equilibrium in an isothermal enclosure is gained in a simple manner by contriving a small aperture in the walls of the enclosure and studying the radiation that escapes through it. If the aperture is sufficiently small, any disturbing effect upon the internal radiation will be negligible. The quantity of light that escapes per second through an orifice of area S is then simply the quantity (S × E) that happens to strike in the same time on any equal surface of the wall.

Naturally there will be no privileged direction for escape.

If therefore, as may easily be done, we look through the aperture, we shall not be able to distinguish any details within the enclosure, the sole impression received being one of a luminous pit, in which nothing definite can be perceived. And the well-known fact is that if one looks through a small opening into a crucible of dazzling molten metal, the surface of the metal cannot be seen. It is not only at low temperatures that nothing can be distinguished within a furnace.

It is, moreover, no more possible at high than at low temperatures to *illuminate* noticeably the inside of the furnace (in such a way as to make its shape visible) by a beam of light passing from the outside in through the small aperture. Such auxiliary light having once entered, it will be dissipated by successive reflections from the walls and will have no chance of getting out again through the aperture in any noticeable quantity. The aperture may be said to be perfectly *black*, if we regard the fact that it reflects none of the light it receives as the essential characteristic of a black body. With regard to the *emissive power* of a black body thus defined, we see that it will be given by the product SE referred to above.

It is not now very difficult to understand how it is possible, by placing two black bodies of this kind face to face, their temperatures being T and t, and one of them functioning as a calorimeter, to measure the excess of energy sent from the hot into the cold source of heat over that sent by the cold into the hot source. In this way it may be proved that the emissive power of a black body is proportional to the fourth power of the absolute temperature T^4 (Stefan's law),

$$E = \sigma T^4,$$

the co-efficient σ being "Stefan's constant."

It is clear that the emissive power increases rapidly as the source of heat gets hotter; when the temperature is doubled, the radiated energy is multiplied 16 times.

The above law has been verified over a wide temperature interval (from the temperature of liquid air to that of melting iron); on theoretical grounds, which are too long to be discussed here, we are inclined to regard it as rigorously exact and not merely an approximation.

The value of Stefan's constant may readily be obtained by making use of the fact that within an enclosure surrounded by melting ice each square centimetre of black surface at the temperature of boiling water loses in one minute very nearly 1 calorie more than it receives (more exactly, 1·05 calories or $1·05 \times 4·18 \times 10^7$ ergs in sixty seconds). In C.G.S. units this gives

$$\frac{1·05 \times 4·18 \times 10^7}{60} = \sigma \ (373^4 - 273^4),$$

or very nearly $6·3 \times 10^{-5}$ for the value of σ.

The density of the light in thermal equilibrium, at the temperature T, being proportional to the emissive power E, is consequently proportional to T^4; or, more precisely, it is equal to $\left(4 \ \frac{\sigma}{c} . T^4\right)$, or $4 \times \frac{6·3 \times 10^{-5}}{3 \times 10^{10}} . T^4$, or $8·4 \times 10^{-15} . T^4$. Though extremely small at the ordinary temperature, it rises very rapidly. Finally, the specific heat of space (the heat required to raise by 1° the temperature of the radiation in 1 cubic centimetre) increases in proportion to the cube of the absolute temperature.[1]

88.—THE COMPOSITION OF THE LIGHT EMITTED BY A BLACK BODY.—The complex light that escapes through a small aperture of area s contrived in an isothermal enclosure may be received on a prism, or, better, on the slit of a spectroscope. It is then seen that such light always behaves as if it were made up by the superposition of a continuous and infinite series of simple monochromatic lights, each producing an image of the slit. Each of these simple monochromatic lights, or, briefly and in the general sense of the word, each of these *simple colours* (visible or invisible) is composed, as shown by Young and Fresnel, of a system of equidistant waves which travel through space at a fixed speed c, which is the same for all kinds of light. A simple

---

[1] It is, in fact, the differential of T with respect to W, being equal to $33·6 \times 10^{-15} \ T^3$; at a temperature of 10,000,000° (the centre of the sun ?) it wou'd be of the order of the specific heat of water at the ordinary temperature.

colour is characterised by its wave length λ, or by its frequency ν (number of waves that pass a given point per second), the product $\nu\lambda$ being equal to the speed c.

By means of screens it is easy to arrange so that not all the energy that issues from the aperture, but only that portion of the energy that corresponds to a narrow band of the spectrum, should enter a "black" receiver acting as a calorimeter; the narrow band will lie between two neighbouring colours having wave lengths λ and λ' and frequencies ν and ν'. This energy tends towards zero as the band becomes narrowed, but the absolute values of its quotient by $(\lambda - \lambda')$ or by $(\nu - \nu')$ tend towards two limits, sI_λ and sI_ν. We will call I_λ the intensity of isothermal radiation (or the intensity of the black body spectrum) for wave length λ, and I_ν the intensity of the same isothermal radiation for the frequency ν.

It is clear that for an infinitely narrow band the products, $I_\lambda(\lambda - \lambda')$ and $I_\nu(\nu - \nu')$ are equal, since they each represent, being defined in the same way, the energy corresponding to that band. Remembering that $\nu\lambda$ (equal to c) is constant, we then see that

$$\lambda I_\lambda = \nu I_\nu.$$

Plotting wave lengths as abscissæ and this intensity as ordinates, a curve will be obtained that shows *the distribution of energy of the spectrum of the black body as a function of the wave length*. In this way it has long been established that the intensity I_λ, which is negligible for the extreme infra-red and extreme ultra-violet, always shows a maximum that varies in position according to the temperature, being displaced towards the region of small wave length (towards the ultra-violet, that is) as the temperature of the black body is raised.

Similarly, taking I_ν as ordinates, we can plot the distribution of the energy of the black body spectrum at each temperature as a function of the frequency. As in the preceding case, the curve starts from zero in the extreme infra-red and returns to zero in the extreme ultra-violet, reaching an intermediate maximum. It must be clearly understood, however, that the two maxima do not correspond to the same colour.[1]

The above are qualitative considerations only. A precise law has been formulated by Wien, who has succeeded in showing that the principles of thermodynamics, although they do not give the actual distribution law required, nevertheless narrow down considerably the number of forms *a priori* possible for it. According to this line of argument, an account of which would lead me into too great a digression, the product of the intensity I by the fifth power of the wave length depends only on the product λT of that wave length by the absolute temperature

$$I_\lambda = \frac{1}{\lambda^5} f(\lambda T),$$

This equation may be written (multiplying and dividing by $\lambda^5 T^5$) :—

$$I_\lambda = T^5 g(\lambda T),$$

the functions f (and g) being as yet indeterminate. From this it follows that if the distribution curve, plotted as a function of the wave length, shows a maximum at a certain temperature, it will show one at all other temperatures and that the position of the maximum will vary inversely with the absolute temperature :

$$\lambda_M T = \lambda'_M T' = \text{constant}.$$

Experiment shows that the product $\lambda_M T$ is constant and that $\lambda_M T = \cdot 29$ very nearly, so that, at $2,900°$ A. (a temperature little lower than that of the electric arc), the maximum intensity corresponds to a wave length of one micron and still lies in the infra-red. At twice that temperature, at about $6,000°$ C. (the temperature of the black body that, put in place of the sun, would send us as much light as the latter), the maximum lies in the yellow.

[1] The wave length for which I_ν is a maximum is very nearly (according to the curves) 1·75 times the wave length at which I_λ is a maximum. If the latter is $0·6\mu$ (yellow, taking in the case of sunlight), the former is then about 1μ, in the infra-red.

If we take the frequency ν instead of the wave length λ as the independent variable, and remembering that λI_λ and νI_ν are equal, we find that Wien's equation takes the form

$$I_\nu = \nu^3 \, F\left(\frac{\nu}{T}\right) = T^3 G\left(\frac{\nu}{T}\right)$$

It remains to determine one of the functions f, g, F or G (which will then give us all the others). Many physicists have attempted unsuccessfully to solve the problem. Planck has proposed an expression that agrees accurately with all measurements [1] in the region between 1,000° and 2,000° C. (absolute) and between 60μ and $\cdot 5\mu$ in wave length. Planck's equation may be written either in the form

$$I_\lambda = \frac{k_1}{\lambda^5} \cdot \frac{1}{e^{\frac{k_2}{\lambda T}} - 1}$$

or

$$I_\nu = \frac{C_1 \nu^3}{e^{\frac{C_2 \nu}{T}} - 1}$$

e being the base of the natural logarithms and k_1, k_2 and C_1, C_2 two pairs of constants, the values of which will be given below.

89.—QUANTA.—The publication (in 1901) of Planck's formula marks an important epoch in the history of physics. It has introduced certain very novel and at first sight very strange ideas into our views on periodic phenomena.

The rays emitted by a black body are, as we have seen, identical with those which, in the isothermal enclosure, traverse a section equal in area to the aperture. From this it follows that in finding the spectral composition of the light emitted, the composition of the light in statistical equilibrium that fills an isothermal enclosure or, more briefly,

[1] Lummer, Kurlbaum, Paschen, Rubens (extreme infra-red), Warburg, and others have carried out these beautiful and difficult measurements.

the composition of the *isothermal radiation*, has at the same time been determined.

In arriving at a theoretical knowledge of this composition we must bear in mind that light waves are electromagnetic.[1] These waves are possibly emitted as the result of oscillatory displacements of electric charges in matter. An electric oscillator (which is a mobile electric charge that may be caused to vibrate by the electric fields due to the waves that successively impinge upon it) can reciprocally and by resonance absorb light having exactly the same period as the oscillator. (*Cf.* Planck's summary of the theory, which is, unfortunately, difficult to follow.)

Let us imagine, within an isothermal enclosure, a large number of identical oscillators vibrating lineally (for example, sodium atoms, such as those regarded as causing the well-known yellow light given by an alcohol flame impregnated with salt). The period of oscillation thus fixed, the light that fills the enclosure must be in statistical equilibrium with these resonators, giving them during the very short period of each oscillation as much energy as it receives from them. If E stands for the mean energy of the oscillators, Planck found that, as a consequence of the laws of thermodynamics, the density w of light for wave length λ is proportional to E, the relationship being expressed more precisely by the equation

$$w_\lambda = \frac{8\pi}{\lambda^4} \cdot E \; ;$$

consequently, in order to reconcile this result with the experimental fact that the radiation density becomes infinitely small for very short wave lengths, it must follow that the mean energy of the oscillators will become extremely small when the frequency becomes very high.

Now oscillators in thermal equilibrium with radiation must also be in thermal equilibrium with any gas that fills the enclosure at the given temperature. In other words, the mean oscillatory energy must be what it would be if it

---

[1] The electric and magnetic fields at a point on the wave front are always in the plane tangential to the wave (light vibrations are transverse) and are perpendicular to each other.

were sustained solely by the impacts of the gaseous molecules. In the case where the oscillatory energy can vary continuously, the kinetic energy of oscillation will, as we have already had occasion to point out, para. 13, be equal in the mean to $\frac{1}{2} \cdot \frac{R}{N} \cdot T$, or to one-third of the kinetic energy of a molecule of the gas : it will, that is to say, be independent of the period. Radiation should therefore be infinite for very small wave lengths, *which is certainly not the case*.

We must therefore assume that the energy of each oscillator varies in a discontinuous manner. Planck supposes that it varies by equal quanta, in such a way that each oscillator always retains a whole number of atoms or *grains of energy*. This granule of energy should be independent of the nature of the oscillator, but should depend on its frequency ν (number of vibrations per second) and be proportional to it (being 10 times greater, for instance, when the frequency becomes 10 times greater) ; it should therefore be equal to $h\nu$, h being a universal constant (Planck's constant).

If we accept these hypotheses, which at first sight appear extraordinary (and which will therefore possess all the more importance if they can be verified), it will no longer be at all accurate to regard the *mean* energy E of a *linear* oscillator as equal to one-third of the energy possessed on the average by a gaseous molecule. Statistical enumeration of all the possible cases shows that in order to arrive at statistical equilibrium through the impacts between gaseous molecules and the oscillators, we must have

$$E = \frac{h\nu}{e^{\frac{Nh\nu}{RT}} - 1},$$

N being Avogadro's number.

On the other hand, Planck shows, though not without some difficulty, that in spite of the hypothesis of discontinuity the density w_λ remains equal to $\frac{8\pi}{\lambda^4}$ E, which gives

LIGHT AND QUANTA 155

for I_λ (which is equal to $w_\lambda \frac{c}{4}$) the expression

$$I_\lambda = \frac{2\pi c^2 h}{\lambda^5} \cdot \frac{1}{e^{\frac{N}{R} \cdot \frac{ch}{\lambda T}} - 1},$$

and, at the same time,

$$I_\nu = \frac{2\pi h}{C^2} \cdot \frac{1}{e^{\frac{N}{R} \cdot \frac{h\nu}{T}} - 1}$$

These formulæ are, in fact, in accordance with experiment, as will be shown later.

The theory I have just outlined has achieved a great success, in that it has led to the discovery of the law that determines the composition of isothermal radiation at each temperature. But a still more striking verification lies in the agreement found between the values already obtained for Avogadro's number and the value that can be deduced from Planck's equation.

90.—THE RADIATION EMITTED BY A BLACK BODY ENABLES US TO DETERMINE THE MOLECULAR MAGNITUDES.—Clearly everything in the above equation is either measurable or known except the number N (which expresses the fact of molecular discontinuity) and the constant h (which expresses the discontinuity of the oscillatory energy). These numbers N and h can therefore be determined if two reliable measurements of the emissive power can be obtained for different values of the wave length λ or the temperature T (it will naturally be better to use in the determination all the available reliable measurements and not two only). Making use of the data that appear most trustworthy at the present moment, we arrive at the following value for h

$$h = 6 \cdot 5 \times 10^{-27}$$

and for N

$$N = 62 \times 10^{22},$$

the probable error being more or less 5 per cent.

The agreement between this value and those already found is indeed marvellous. And at the same time we have acquired yet another means for determining accurately the molecular magnitudes.

In spite of the importance of these results, we cannot conceal the fact that Planck's theory presents great difficulties. It will be (and already has been) profoundly modified, and we shall certainly see Planck's postulates replaced by others more comprehensible and in more accurate agreement with experiment (Bohr).

Extension of the Theory of Quanta.

91.—The Specific Heat of Solids.—By a bold extension of Planck's idea Einstein has succeeded in accounting for the influence of temperature on the specific heat of solids. His theory, to which allusion has already been made (para. 44), depends upon the assumption that each atom in a solid body is urged towards its position of equilibrium by elastic forces in such a way that, if it be slightly displaced, it will vibrate with a fixed period. As a matter of fact, since neighbouring atoms also vibrate, the frequency will not be rigidly fixed, and we ought rather to consider a series of frequencies more analogous to a band than to a spectral line. Nevertheless, as a first approximation we may confine ourselves to the consideration of the case of a single frequency.

With this limitation, Einstein supposes that although the oscillator set up by each atom is not necessarily an electrical one, its energy must be a whole number multiple of $h\nu$ as with Planck's oscillators. Its mean energy at any temperature has therefore the value

$$\frac{3h\nu}{e^{\frac{N}{RT} \cdot h\nu} - 1},$$

with reference, as has been pointed out above, to an oscillator capable of undergoing displacement in all directions. The energy contained in a gramme atom will be N times greater and the increase of this energy per degree, or the specific heat of the gramme atom, can be calculated.¹ The expression found in this way for the specific heat tends towards zero, in agreement with Nernst's results, as the temperature falls, and towards 3R or 6 calories as it rises, in agreement with Dulong and Petit's law (the latter limit is reached the more quickly the smaller the value of the characteristic frequency ν). In the interval between these limits the above expression represents in a remarkable manner the run of the specific heats, though not without systematic error explicable by the approximations employed (we have noticed that the frequency cannot be defined satisfactorily). It also defines the frequency ν of the atomic vibration, if it is unknown.

It is worthy of notice that frequencies calculated in this way agree with those to be expected from the consideration of other phenomena. The absorption of light of long wave length by bodies such as quartz or potassium chloride (Rubens' experiments) is a case in point. This kind of absorption, as well as " metallic " reflection, is explicable if the light is in resonance with the atoms of the body and consequently possesses a frequency deducible from the latter's specific heat. This is found approximately to be the case (Nernst).

At the same time it is conceivable (Einstein) that the elastic properties of solid bodies might provide a means for predicting the frequency of the vibrations of an atom displaced from its equilibrium position. An approximate calculation has been made by Einstein with reference to *compressibility*; applied to silver the predicted value for the atomic frequency is 4×10^{12}, that obtained from its specific heats being 4.5×10^{12}. I must content myself with these brief allusions and refer to the splendid work of Nernst, Rubens, and Lindemann ² for more ample details.

92.—DISCONTINUITY IN ROTATIONAL VELOCITY.—If we remember that we have already been forced to assume, with Nernst (para. 45), that the rotational energy of a molecule

---

[1] It will be simply the differential with respect to temperature of the energy contained in a gramme atom.
[2] The latter deduced the proper frequency from the melting point of the solid; he supposes that a body liquefies when the amplitude of the atomic oscillation becomes sensibly equal to the mean distance between atoms.

varies discontinuously, we shall perhaps be readier to extend, keeping the same value for the universal constant h, the law of discontinuity which holds for the energy of oscillators to the case of rotation (molecular). There is, indeed, analogy of a certain kind, since periodicity is characteristic of both cases. An obvious difference is that an oscillator has a well-defined proper period, whereas so long as a ball is at rest, it is not possible to predict a definite period of rotation for it. However, generalising Planck's postulates, we may say with Bjerrum :

When a body rotates at the rate of ν revolutions per second, its energy is equivalent to a whole number of times the product $h\nu$.

Since $2\pi\nu$ is the angular velocity of rotation (angle described per second), the kinetic energy of rotation is furthermore equal to the product $\frac{1}{2}I(2\pi\nu)^2$, where I stands for the moment of inertia [1] of the body (about its axis of rotation). Hence it follows that, p being a whole number,

$$\frac{1}{2} I \cdot 4\pi^2\nu^2 = p \cdot \nu \cdot h,$$

or
$$\nu = p \cdot \frac{h}{2\pi^2 I},$$

so that the number of revolutions per second must necessarily be either once, twice, or 3 times a certain value t equal to $\left(\frac{h}{2\pi^2 I}\right)$. Intermediate speeds of rotation should be impossible.

A more searching analysis of the relation between a rigid rotating body and a linear oscillator has led Sommerfeld to modify slightly the above result ; he finds that *twice* the kinetic energy of rotation should have values that are whole number multiples of $h\nu$. This introduces no essential change into the result, which is the implication of discontinuity in the speeds of rotation. The only alteration is that the value of t (of which the number of rotations per second is a whole

---

[1] We know that the rotational energy of a solid revolving with an angular velocity ω is $\frac{1}{2}I\omega^2$ (which may be used to define its moment of inertia).

LIGHT AND QUANTA 159

number multiple) becomes $\frac{h}{4\pi^2 I}$. Intermediate speeds of rotation remain impossible.

93.—UNSTABLE ROTATION.—The above result is surprising ; it appears, moreover, to be inconceivable that the number of revolutions can pass from the value t to the value $2t$ or $3t$ without taking up the intermediate values. I would suggest that the intermediate velocities are *unstable*, and that when, for instance, the body while rotating receives an impulse that communicates to it an angular velocity corresponding to 3·5 times t revolutions per second some *effect due to friction or radiation as yet unknown at once operates to reduce the number of revolutions per second to exactly 3 times* t, *after which the rotation can persist indefinitely without loss of energy.* The result of this will be that, out of a large number of molecules, very few will be in the unstable condition, and we may take it as a first approximation that, for any one molecule taken at random, the rotation in one second is either no revolutions or t or $2t$ or $3t$, etc. The occasional molecules having rotational energy in process of change may be neglected, just as we may neglect the few molecules in a gas that are actually undergoing impact and whose energy is in process of changing.

94.—THE MATERIAL PART OF AN ATOM IS CONCENTRATED ENTIRELY AT ITS CENTRE.—We are now possibly in a position to understand why the molecules of a monatomic gas (such as argon) do not produce rotation when they impinge (or, more exactly, why they communicate no rotational energy to each other), with the result that the specific heat c of the gas is equal to 3 calories (para. 39). If the material part of the atom is concentrated closely about its centre, its moment of inertia will be very small, its minimum possible rotation (its frequency ν being $\frac{h}{4\pi^2 I}$) will be extremely rapid and the quantum h^ν of rotational energy consequently will be large. If this quantum is large by comparison with the energy of translation possessed on the average by the molecules (at the temperatures we have available), it will practically never happen that a molecule that strikes another molecule will be able to communicate to it even the minimum rotation ; and, conversely, a molecule possessing that rotation will have every chance of losing it during an impact. In short, at any particular instant, rotating molecules will be extremely few in number.

Since argon, to take a particular case, retains its specific heat 3 up to about 3,000° C., it follows that even at that high temperature the molecular translational energy is still well below the quantum of energy corresponding to the minimum possible rotation. Let us assume that the translational energy is less than half the quantum, which is certainly a very low estimate. Further, since it is proportional to the absolute temperature, it will be approximately 10 times greater than at the ordinary temperature and hence very nearly equal to $\frac{1}{2} \times 10^{-12}$; the quantum $h\nu$ being expressible in the form $\frac{h^2}{4\pi^2 I}$, this gives us

$$\frac{1}{2} \times 10^{-12} < \frac{1}{2} \times \frac{h^2}{4\pi^2 I}.$$

Substituting for h its value 6×10^{-27}, it becomes possible to deduce from this inequality some interesting results with regard to the frequency and moment of inertia.

In the first place, if $h\nu$ is greater than $\frac{1}{2} \times 10^{-12}$, we see at once that ν is certainly higher than 10^{14}:

The slowest stable speed of rotation corresponds to more than 10,000,000,000 revolutions in one hundred-thousandth of a second.

With regard to the moment of inertia, we see that it is less than 10^{-42}. If the mass m of the argon atom (equal to 40 times the mass $1 \cdot 5 \times 10^{-24}$ of the hydrogen atom) occupies a sphere of radius r with uniform density, its moment of inertia would be $\frac{2}{5} m r^2$, which would give

$$r < 2 \times 10^{-10},$$

and, more generally, we cannot imagine a distribution of the atomic mass compatible with the smallness required for the moment of inertia, and such that any appreciable part of that mass should be at a distance from the centre well above 10^{-10} centimetres.

The material part of the atom is thus condensed almost entirely (if not completely) into a *kernel* the dimensions of which are small in comparison with the diameter derived from the kinetic theory (for argon about 3×10^{-8} centimetres).

The extraordinarily dense kernel would then have a volume at least 10^6 times smaller than the apparent volume of the atom (defined as the distance between the centres on impact or by their distance apart in the crystalline state at very low temperatures). In fact, the degree of condensation is probably much greater even than this, and we shall see that the rarity of the durations undergone by a rays indicates an actual volume of about 10^{12} times smaller than the apparent volume of the atom. This is very nearly the ratio between the volume of the sun and that of a sphere that includes the orbit of Neptune.

In other words, if we could examine a solid body with a microscope magnifying 3×10^{10} times the body would appear to us to be composed of extremely dense granules about 2 millimetres in diameter, with a mean distance of about 20 metres between them.

Finally, we shall see that an extremely small fraction of the atomic mass, composed of negative electrons gravitating around the nucleus, remains outside the latter.

Matter is porous and discontinuous to an extent far beyond our expectation.

The radius of protection, or distance between centres at the moment of impact, may be defined, as we have already suggested, as the distance at which the material of the atom exerts an enormous repulsive force upon the material part of another atom. We shall see, when discussing the rapid positive rays, that at still smaller distances the force of repulsion becomes feeble or fails altogether. In other words, every atom is condensed at the centre of a casing or sheath, which is of vast dimensions relatively to the atom itself and which protects the latter from the approach of other atoms. We shall see later that this sheath is probably composed of electrons, the existence of which on the periphery of the atom we have just indicated.

95.—THE ROTATION QUANTUM OF A POLYATOMIC MOLECULE : THE DISTRIBUTION OF MATTER WITHIN THE MOLECULE.—We can now understand why even a molecule may cease to spin at very low temperatures, although its moment of inertia is much greater than that of a single atom. The only necessary condition is that the molecular energy due to agitation should be small in comparison with the rotation quantum $\frac{h^2}{4\pi^2 I}$ for the molecule. Naturally this will occur sooner the smaller the moment of inertia of the molecule, and we can understand why such a state of affairs has been realised as yet with hydrogen alone (para. 45).

Let d be the distance between the centres of two hydrogen atoms making up a molecule H_2. Their masses are concentrated at the two extremities of d, and the moment of inertia I about an axis passing through the centre of gravity of the molecule and perpendicular to the line joining the centres will be

$$2 \times 1\cdot 4 \times 10^{-24} \left(\frac{d}{2}\right)^2.$$

At about 30° absolute (at which temperature the specific heat is very nearly 3) the quantum $\frac{h^2}{4\pi^2 I}$ will certainly be greater than twice the energy due to molecular agitation, which, at this temperature, is very nearly $\frac{1}{2} \times 10^{-14}$. It follows from this that the distance between the centres is certainly less than $1\cdot 2 \times 10^{-8}$. We shall readily accept this upper limit when we remember that $2\cdot 1 \times 10^{-8}$ was found for the diameter of impact of the hydrogen molecule.

A somewhat more accurate calculation is possible if the small difference between the actual specific heat at say, 50° and 3 calories is known. In this way it was found that d is roughly 10^{-8} centimetres.

We have seen how small in reality is the space occupied by

LIGHT AND QUANTA

the atoms within the molecular edifice. It would be of the greatest interest to know the distribution of the field of force about each atom and particularly to gain precise ideas as to the nature of the chemical bonds or valencies.

In this connection I should like to add a remark with reference to the *strength of the valency bond*. When, at about 2,000° C., the dumb-bell-like hydrogen molecule is spinning without rupture perpendicularly to its axis with a frequency but little less than a hundred thousand milliards of revolutions per second, it is obvious that the bond or union between the atoms must be resisting the centrifugal force. A union that would give the same strength to a dumb-bell would have a tenacity at least 1,000 times that of steel.

96.—LIGHT MAY POSSIBLY BE THE CAUSE OF MOLECULAR DISSOCIATION.—I have indicated (para. 84) the possibility of a kinetic theory of chemical reaction. I should like to point out that light may possibly play a capital *rôle* in the mechanism of chemical change.

This appears to me to be proved by a law [1] that is quite generally recognised, without, I think, a sufficient appreciation of its really surprising molecular interpretation, which may perhaps show it to be the fundamental law of chemical mechanics (since all chemical equilibria presuppose certain molecular dissociations).

According to this law, the rate of dissociation at constant temperature, in unit volume of a gas A, for a reaction of the following kind :

$$A \longrightarrow A' + A'',$$

is proportional to the concentration of the gas A and cannot be altered by the addition of other gases to the reacting system.

In other words, for a given mass of the substance A, the proportion transformed per second is independent of the dilution ; if the given mass occupies 10 times more space, its concentration then being 10 times less, then 10 times

---

[1] As a matter of fact, experiments referring to *gases* are few in number.

less will be transformed per litre, or just as much in all as before dilution. Thus, contrary to what might have been expected, *the number of impacts has no influence on the rate of dissociation*. Out of the given N molecules of the gas A, always the same number will decompose per second (at a given temperature) whether the gas is relatively concentrated or mixed with other gases (in which case impacts will be frequent), or whether it is dilute (when impacts will be rare).

It seems to me that, for any given molecule, the probable value for the time that must elaspe before, under the sole influence of impacts, a certain *fragile* condition will be reached must be smaller the more often the molecule receives impacts per second. Further, supposing this fragile state to have been reached, the probable value for the time required for a molecule to receive the kind of impact capable of rupturing it must again be shorter the more frequent the impacts. For this double reason, if rupture is to be produced by molecular impact, it should occur more frequently (and dissociation should therefore become more rapid) as the concentration of the gas increases.

Since this is not the case, dissociation cannot be caused by impact. Molecules do not decompose by striking against each other, and we may say : *The probability that any molecule will be ruptured does not depend upon the number of impacts it receives.*

Since, however, the rate of dissociation depends largely on the temperature, we are reminded that temperature exerts its influence by radiation as well as through molecular impact, and are faced with the suggestion that the cause of dissociation lies in the visible and invisible *light* that fills, under stationary conditions, the isothermal enclosure wherein the molecules of the gases under consideration are moving.

The essential mechanism of all chemical reaction is therefore to be sought in the action of light upon atoms.

96A.—RADIO-CHEMISTRY.—Since the above lines were written (1912) a number of attempts have been made to develop a satisfactory theory of radio-chemistry.[1] I shall

[1] It will have been seen above how the idea of the part played by light was

indicate here how it seems to me possible to explain the law of Planck and quanta in terms of the discontinuous properties of matter.

Let us take the most general chemical reaction

$$n_1A_1 + n_2A_2 + \ldots \to n'_1A'_1 + n'_2A'_2 + \ldots$$

or, in a still more condensed form, $A \to A'$. On the molecular scale the reaction would consist in the sudden appearance of a molecular grouping A' at a spot where there had been a grouping A.

It is universally admitted (law of mass action) that the reaction velocity $v_{AA'}$ (the quotient of the number of groupings A' which appear at the expense of A_1 per unit volume per second divided by the Avogadro number N), expressed as a function of the concentration C is, at a given temperature,

$$v_{AA'} = kC_{A_1}^{n_1} \cdot C_{A_2}^{n_2} \ldots \ldots$$

Let us deal with the apparently most simple reaction, that of the rearrangement of the matter of a molecule: $A_1 \to A'_1$. We suppose that at a definite temperature the reaction velocity tends to approach zero when the temperature is lowered; every chemical reaction so far studied teaches that. At absolute zero, therefore, the molecule A_1 would subsist indefinitely, without any "spontaneous" passing into the state A'_1. Similarly, the molecule A'_1 would subsist indefinitely. In short, the two states, A_1 and A'_1, are *stable*.

forced upon me as a logical necessity. I believed then that the idea was entirely new. As a matter of fact, Trautz had previously suggested that ordinary reactions are possibly brought about by infra-red light.

The outbreak of the war interrupted and isolated research. On once more taking up my work, I developed a general theory of radio-chemistry ("Matière et Lumière," Ann. de Phys., 1919). I was then ignorant of the fact that, during the war, W. C. McC. Lewis, himself unaware of my earlier publication, had published some important papers on the same lines ("Studies in Catalysis," Trans. Chem. Soc., 1916–1918). I also discovered later that Trautz, in ignorance of both of the above contributions to the subject, had developed similar ideas (though not without some obscurities) in a long paper on thermal and photochemical phenomena (Zeits. für Anorg. Chem., 1918). Finally Sheppard (Photo-chemistry) had come very near the essential idea.

Such coincidences are well known to be frequent in the history of science.

We may presume that each of these states is in reality a stationary, quasi-periodic condition of electronic movements. We must insist, however, that this *rotation of charged centres is not accompanied by any radiation;* unless this be so the internal energy would continuously diminish, and there would not be any state A that we could define.

Let the letter L designate *in quantity and in nature* the "radiation," the absorption of which can determine the disappearance of the stable state A_1 which is followed by the appearance of the stable state A'_1. This same radiation L will be released and emitted, in the inverse reaction $A_1 \leftarrow A'_1$. Let L' similarly signify, in quantity and in nature, the radiation, the absorption of which causes this reversed reaction; L' must then have been emitted at the time of the direct reaction. Instead of expressing by the symbol $A_1 \rightarrow A'_1$ the transformation of an isolated molecule in a vacuum, a transformation which hence implies absorption and then emission of radiation, we shall make use of the symbol

$$\overrightarrow{L} + A_1 \rightarrow A'_1 + \overrightarrow{L'}.$$

(The inverse transformation would be expressed by inverting the arrows.)

If the molecule can completely be isolated at the moment of its remaking, it can only take up energy from the radiation by which it is surrounded, and it cannot lose energy except in the form of radiation. If, therefore, $U_{AA'}$ designates the gain of internal energy realised when A_1 has been transformed into A'_1, it follows necessarily that

$$U_{AA'} = L - L'.$$

This variation of energy will be negative or positive according as the reaction is exothermic or endothermic, but it will always appear in the form of a difference of two finite positive quantities. We must always commence by supplying energy even if, in the end, we should receive much more energy. (In order to explode a shell which is, in a sense, a stable thing, it is not sufficient to caress the shell; a definite very notable, amount of work has to be done on the point of the shell.)

Critical Molecules, Spontaneously Transformable.—Neither of the two energies of the radiations L and L' can be of value zero, because the two states A and A' were stable.

On further reflection we shall, I think, regard it as probable that a state a, spontaneously transformable, will be obtained when the matter, first in the state A_1, has acquired the energy L. It does not remain in that state; after the absorption it transforms itself spontaneously into the state A' while emitting L' (or it will return spontaneously to the state A while re-emitting L, in which case there will have been no reaction in the end). We say that this state a is a *critical state*.[1]

A critical state of the same internal energy as a, which must be identical with a_1, will also be attained when A' has conversely absorbed the energy L'. It appears to me reasonable to suppose that the probability of the transition from the critical state into either of the stable states has a well-defined value such that there would be a definite *average life* for the critical state. But these probabilities are *internal*, and the spontaneous fall towards one of the stable states would be produced even in the vacuum, at absolute zero, as a consequence of internal accidents concerning which we know nothing.

That critical state does not directly admit of observation, because its average life θ is negligible by comparison with the lives Θ and Θ' of the stable states A and A'; thus the concentration C_a is negligible compared to C_A and $C_{A'}$. I presume, however, that we shall obtain evidence of the existence of critical molecules. Perhaps that has already been done by the recent experiments of Wood,[2] showing that in mercury vapour the absorption of a certain ultra-violet light and the subsequent emission of a fluorescence are

---

[1] René Marcelin (Ann. de Phys., 1915), from considerations of mechanical statistics, and Lewis (Trans. Chem. Soc., 1917), by a somewhat different method, had already been led to the assumption of a critical state, from which the matter passes spontaneously into one of the states, A or A'.

[2] Proc. Roy. Soc., 1921.

separated by a definite interval, of the order of $\frac{1}{15000}$ of a second.

The following diagram may elucidate these conceptions. Different levels are to correspond to the different quantities of internal energy which the matter studied can contain, a low level indicating small energy. The exothermic system A corresponds to the lowest level, and the critical state a, from which the matter may fall back towards A or A', to the highest level. The radiation L is absorbed in the transformation Aa, or liberated in the spontaneous transformation

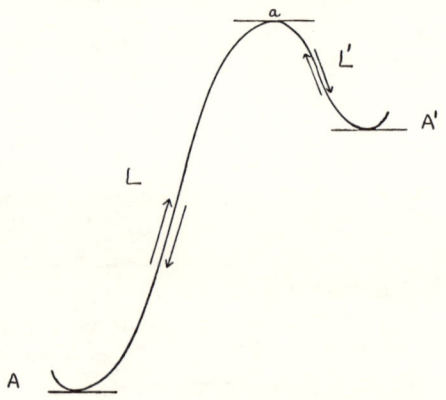

Fig. 12.

aA; similarly L' is absorbed along A'a or liberated in the spontaneous transformation aA'. The matter cannot exist in any intermediate state whatever between a and A or between a and A', without either emitting or absorbing. But in a itself, it does not absorb any more than it does in A or A'. The state a is a critical non-radiating state, just as A and A' are stable non-radiating states.

In short, there are, for the matter of a molecule, generally several values of the internal energy for which a stationary non-radiating state is realised. There are the stable states which persist indefinitely unless external energy be brought up, and the critical states from which, apart from internal accidents, the matter passes *explosively* into a stable state. There will exist, for each stable state, within a given radiation, a definite probability for the absorption of the energy which will transform it into one of the critical states of superior energy. And for each critical state there will be a spontaneous explosive fall back into one of the critical or stable states of inferior energy, independent of external conditions,· but with a definite internal probability, and hence in definite proportions.

The transition from A into A' may, *e.g.*, be effected either *via* a critical state a_1, with absorption of L_1 and liberation of L'_1, or *via* another critical state a_2, with absorption of L_2 and liberation of L'_2. That will make two active radiations, subject to the necessary condition that $(L_1 - L'_1)$ be equal to $(L_2 - L'_2)$. It appears that the active radiation may be complex.

Influence of the Temperature.—The velocity constant is known to become larger as the temperature rises. We easily understand why. Let us again take the simple case of a tautomeric reaction. The internal readjustment takes place because the exciting radiation is always present in the isothermal radiation in which the molecule is immersed. If the probability of the readjustment increases with the temperature, it is because the intensity of this radiation increases. The temperature intervenes only as an intermediate variable determining the fundamental variable, the intensity of the exciting radiation. Provided that this intensity take the same value, we should find the same reaction velocity, whatever be otherwise the composition of the radiation in which the molecule is immersed.

Arrhenius' Formula.—On the other hand, Arrhenius has succeeded in expressing the influence of the temperature on the constant of the velocity by an equation which may be written

$$k = se^{-a/RT},$$

where s and a designate two constants which are fixed for the reaction. The constant k' of the velocity of the inverse reaction will be given by an analogous formula, and the ratio of the two constants is then

$$k'/k = e^{\frac{a-a'}{RT}} \cdot s'/s.$$

Chemical equilibrium being established when the velocities kC_A and $k'C_{A'}$ of the inverse reactions become equal, the equilibrium concentrations will have the ratio

$$C/C' = k'/k = e^{\frac{a-a'}{RT}} \cdot s'/s.$$

Moreover, for the above transformation in statistical equilibrium, the ratio between the two states A and A_1 should be very nearly $e^{U_{AA'}/RT}$. This was pointed out by Marcelin,[1] who obtained a statistical formula of which the law of distribution of molecular velocities (Maxwell) is a special case, and which is now regarded as being of general validity. From this it follows that $U_{AA'}$ is equal to $(a - a')$. We thus obtain the physical significance of a (or a'): a measures the gain of internal energy between the state A and the critical state a—a gain which one may, with Lewis, call the " critical increment " in starting from A.

Planck's Law.—I will now show how, if we assume that radiation determines the reactions, the formula of Arrhenius can give information concerning the isothermal radiation and can introduce the quantum relation. We have shown that the variation of the internal energy $U_{AA'}$ is equal to $(L - L')$; we have just found that it is equal to $(a - a')$. The meaning of the two terms a and L may induce us henceforth to regard the double equality, $L = a$ and $L' = a'$, as very probable. Let us consider the simple case in which the active radiation is monochromatic [2] and of frequency ν. Let I_ν be the intensity of this radiation in the isothermal radiations; the velocity constant k is a function $f(I_\nu)$, and it also obeys Arrhenius' law: $f(I_\nu) = se^{-a/RT}$. We should like to say that the velocity is proportional to the intensity; this holds certainly (as the reactions which are already classed as photo-chemical show) for feeble intensities. But the formula just mentioned indicates the

---

[1] Killed in action. " Cinétique Physicochimique," Ann. de Phys., 1915.
[2] I am inclined to believe that, on the molecular scale, this is always the case. The active radiation is really split up, in a continuous or discontinuous domain of radiations, into monochromatic radiations which are separately active for one or other molecule.

limiting value s for large values of I_ν.[1] The function $f(I_\nu) = sI_\nu/(B + I_\nu)$, where B is a constant, appears to me to be the simplest of those which satisfy both the conditions. We have then $I_\nu/(B + I_\nu) = e^{-a/RT}$, which gives us the intensity I_ν, that is to say, the distribution of the energy in the spectrum of the isothermal radiation by the formula $I_\nu = B/(e^{a/RT} - 1)$. On the other hand, we have the thermodynamical equation of Wien: $I_\nu = \nu^3 F(\nu/T)$, where F is a universal function *in vacuo*. Combination of the two equations leads to $B = C\nu^3$, and $a = H\nu$, C and H being two universal constants, so that $I_\nu = C\nu^3/(e^{H\nu/RT} - 1)$. This is the formula which Planck found by an obscure and difficult method. I do not think that there is any more simple way of introducing [2] this formula.

Quanta. Law of Bohr.—We have at the same time introduced the important idea of the quantum. We have, in fact, found that the energy L or a, retained by molecules which pass from one definite state to another definite state, is $H\nu$ per gramme molecule. Per molecule the value is $H\nu/N$ or $h\nu$, h being the universal constant, approximately equal to $6 \cdot 5 \times 10^{-27}$.

When, therefore, a radiation of frequency ν causes a molecule, isolated in space, to pass from one stationary state to another stationary state, the internal energy of this molecule is increased by the quantum $h\nu$.

Reciprocally (according to the postulate of reversibility), when a molecule passes from a critical state to a stationary state of smaller energy, and when this takes place in a vacuum the molecule gives up the lost energy in the form of mono,

---

[1] We can understand that as the illumination becomes more and more intense, the fragility grows less and less rapidly. An " efficient " encounter between radiation and molecule demands, indeed, in my opinion, simultaneously (1) that the waves which strike the molecule have, with respect to the molecule, certain dispositions of orientation or of phase, the probability of which seems independent of the intensity, and hence of the temperature, and (2) that these waves continue to carry sufficient energy throughout the duration of this favourable disposition.

This second condition will be realised more and more frequently as the temperature is raised, and will finally always be fulfilled in practice ; thus the fragility will tend towards a final limit.

[2] Unfortunately this deduction does not explain the signification of the constant C, equal to $2\pi/c^2 \cdot H/N$.

chromatic radiation, the frequency of which is given by the quotient of the energy lost by the universal constant h.

This law, which is of obvious importance, is the generalisation of one of the hypotheses by means of which Bohr succeeded, in 1913 (Phil. Mag.), in explaining the emission spectrum of hydrogen. It has been seen that it can be readily introduced.

CHAPTER VII

THE ATOM OF ELECTRICITY

WE have seen that the properties of electrolytes suggest the existence of an indivisible electric charge and that every ion either carries that charge or some whole number multiple of it. But we have not as yet attempted the *direct* measurement of this elementary charge and have merely calculated its value by dividing by Avogadro's number N the electric charge (1 faraday) carried during electrolysis by a monovalent gramme ion.

Now the direct measurement of very small charges, which up to the present has not been successfully accomplished in liquids, is found to be easy with gases, and has in fact shown that such charges are always whole number multiples of the same quantity of electricity, which has a value agreeing with that already calculated. Experiments that I am about to describe have proved the discontinuous structure of electricity and have provided yet another means for obtaining the molecular magnitudes.

97.—KATHODE RAYS AND X RAYS: THE IONISATION OF GASES.—Since the time of Hittorf (1869) it has been known that when an electric discharge passes through a rarefied gas, the kathode emits rays which show their trajectory by a feeble luminescence of the residual gas, excite a beautiful fluorescence on the glass walls at which they are arrested, and which are deviated by magnets. If, for instance, they are directed at right angles to a uniform magnetic field, their trajectory becomes circular and perpendicular to the field.

As early as 1886 Sir W. Crookes supposed that these *kathode rays* are negatively electrified projectiles which, issuing from the kathode and being repelled by it, have acquired an enormous velocity. But neither he nor Hertz was able to prove the existence of this electrification, and a wave theory was favoured for some time, although Hertz had discovered that the rays can pass through thin plates several microns thick and Lenard had shown that it is possible to allow them to escape from the tube wherein the discharge takes place through a thin metallic plate strong enough to support the pressure of the atmosphere. (They can then be studied in the air, into which they diffuse and are stopped after a path of a few centimetres.)

A decided reversion to the emission theory put forward by Crookes occurred, however, when it was proved [1] that the kathode rays always carry negative electricity along with them. It is absolutely impossible to separate this electricity from the rays, even by making them pass through thin metallic leaf.

Finally, we may remark that any obstacle struck by the kathode rays emits X rays, the discovery of which by Roentgen (1895) marked the commencement of a new era in physics.

Like kathode rays, X rays excite fluorescence of various kinds and affect photographic plates. They differ profoundly from kathode rays in that they carry no electric charge and consequently are not deviated either by electrified bodies or by magnets. It is well known that they possess a very considerable penetrating power and that they cannot be reflected, refracted or diffracted, so that they are of much shorter wave length than the most extreme ultraviolet light ($\cdot 1\mu$) yet studied.[2]

But according to the brilliant suggestion of Laue, crystals, being composed (according to the theory of Hauy and Bravais) of molecules distributed in a rigorously periodic manner, should behave as three-dimensional diffraction gratings ; the spacings of the fundamental intervals are

---

[1] Jean Perrin (Comptes Rendus, 1895 ; Ann. de Ch. et Phys., 1897). I showed that the rays carry negative electricity with them into a completely closed metallic box and that they are, moreover, deviated in an electric field.

[2] The work of Laue, Bragg, etc., has shown that the X ray spectrum extends form $\lambda = \cdot 00084\mu$ to $\lambda = \cdot 000056\mu$ (cf. Moseley, Phil. Mag., April, 1914).—Tr.

THE ATOM OF ELECTRICITY 175

of the order of the molecular diameters, and should be able to reveal wave lengths of the same order of magnitude.

If, therefore, a beam of X rays made up of a series of " colours " impinges on a crystal, it should give rise to a series of diffracted monochromatic rays in directions determined by the symmetry and orientation of the crystal and by the distribution of its molecules. Experiment has completely verified these predictions (1913), at the same time revealing to us the structure of crystals [1] and the wave lengths of X rays, which, in general, are about 1,000 times shorter than visible light. They are, on the one hand, " harder " (more penetrating) the shorter their wave length ; on the other hand, there are rays which correspond to ordinary ultra-violet light in being completely absorbed by thin layers of gases.

IONISATION.—It was very soon noticed that X rays " discharge electrified bodies." A careful analysis of the phenomenon [2] showed that the rays produce, in the gases they pass through, nuclei charged with positive or negative electricity, or mobile ions, which soon recombine in the absence of an electric field, but which move under the influence of such a field in opposite directions along the lines of force until stopped by a conductor, which discharges them (thus enabling the degree of ionisation of the gas to be determined), or by a non-conductor, which they charge. Once the oppositely charged ions have been carried by this two-fold motion into different regions of the gas, they escape recombination and the two electrified gaseous masses thus obtained can be manipulated at leisure.

In the same way, by their ionising effect on gases, other forms of radiation were soon afterwards detected (extreme ultra-violet rays, Lenard's kathode rays, the α, β, and γ rays of radioactive substances) that " discharge " electrified

---

[1] W. H. Bragg and W. L. Bragg have obtained remarkable results in this field.

[2] Jean Perrin : " Mécanisme de la décharge des corps électrisés par les rayons X "; " Eclairage électrique," June, 1896; Comptes Rendus, August, 1896 ; Ann. de Ch. et Phys., August, 1897. Sir J. J. Thomson and Rutherford have arrived at the same conclusions, from their side, as the result of quite different experiments. Righi has also reached the same position.

bodies situated in gases, when they cut the lines of force emanating from those bodies. The gases issuing from flames are also ionised, and we may reckon them as conductors as long as any ionisation persists.

98.—THE CHARGES SET FREE DURING THE IONISATION OF GASES ARE EQUAL IN VALUE TO THOSE CARRIED BY MONOVALENT IONS DURING ELECTROLYSIS.—As yet we know nothing as to the magnitude of the charges separated during the ionisation of a gas, or whether they bear any relation to the ions concerned in electrolysis.

That the elementary charges are the same in the two cases [1] was first shown by Townsend. Thus let e' be the charge on an ion, situated in a gas of viscosity ξ. Under the influence of a field H this ion will be set in motion, and, being continuously checked by the impacts it receives, it will be displaced with a uniform motion (on our dimensional scale) at a velocity u such that

$$\mathrm{H}e' = \mathrm{A}u,$$

the coefficient of friction A no longer having the value $6\pi a\xi$ that it takes (para. 60) for a relatively large spherule; it is constant, however, which is all we require. As a matter of fact, u can be measured (Rutherford), and it can be shown that the quotient $\dfrac{u}{\mathrm{H}}$, which may be regarded as a *mobility*, is constant, and furthermore not the same for each of the two kinds of ion produced. This mobility corresponds roughly to a velocity of 1 centimetre per second in a field of 1 volt per centimetre.

After the separation of the two kinds of ions by the electric field, two gaseous masses are obtained in which ions of the same kind only are to be found. These ions are in a state of agitation and diffuse just like the molecules of a very attenuated gas scattered throughout a non-ionised [2] gaseous medium.

Hence, making use of Einstein's argument (para. 70), we

---

[1] Phil. Trans., 1900

[2] Neglecting the extremely weak repulsive action that tends to drive these mobile charges towards the periphery of the enclosure.

THE ATOM OF ELECTRICITY 177

find for the value D for the diffusion coefficient of the ions under consideration

$$D = \frac{RT}{N} \cdot \frac{1}{A},$$

that is, since A is equal to $\frac{He'}{u}$,

$$Ne' = \frac{RT}{D} \cdot \frac{u}{H}.$$

This is Townsend's equation (obtained by him, as a matter of fact, by a different method).

To obtain the product Ne' we have only, since the mobility $\frac{u}{H}$ is known, to measure the diffusion coefficient D. This Townsend has done. He found, for various gases and the various kinds of ionising radiation, that the value of the product Ne' is in the neighbourhood of 29×10^{13}, which is the value obtained for Ne from electrolysis.[1]

A later verification, in connection with the very interesting case of the ions in flames, follows from Moreau's[2] experiments on the mobility and diffusion of such ions. The value 30.5×10^{13} is obtained for Ne, which is equal to the value given by electrolysis to within 5 per cent.

Remembering that, on account of the irregularity of molecular motion, the coefficient of diffusion is always equal to half the quotient $\frac{X^2}{t}$ characteristic of the agitation (para. 71), we can re-write Townsend's equation in the form

$$Ne' = 2\,RT \cdot \frac{t}{X^2} \cdot \frac{u}{H},$$

which, though without interest in connection with the invisible ions dealt with by Townsend in his experiments, becomes the most interesting form in the case of *large ions* (charged powders), if their displacements can be measured.

---

[1] A small proportion of other kinds of charges (polyvalent ions, for instance) may have escaped observation; the degree of uncertainty attached to the measurements appears to be about 10 per cent.

[2] Comptes Rendus, 1909.

This has actually been done by de Broglie, using air charged with tobacco smoke.[1] In his apparatus the air and smoke is blown into a small box maintained at a constant temperature, and luminous rays are caused to converge into it from a powerful source. At right angles to these rays a microscope is fixed, which resolves the smoke into globules that look like brilliant points of light and are agitated by a very active Brownian movement. If now an electric field is produced at right angles to the microscope, the globules are seen to be of three kinds. The first kind move in the direction of the field and are therefore positively charged; others move in the opposite direction and are therefore negative. Finally, there is a third group, which continue their agitation without changing their position and are therefore neutral. In this very striking manner *large gaseous ions* were for the first time made visible.

De Broglie has carried out a large number of measurements of X and of u for ultra-microscopic globules of very nearly the same brightness (and hence of about the same size). The mean of these experiments gives the value $31 \cdot 5 \times 10^{12}$ for Ne'; we thus obtain, with a degree of accuracy equal to that obtaining in Townsend's experiments, a value equal to the product Ne given by electrolysis.

More recently Weiss (Prague) has found the same value of Ne' for the charges carried by the ultra-microscopic particles that occur in a spark passing between metallic electrodes.[2] But, instead of taking the means of isolated readings relative to different grains, he recorded for each grain enough readings to obtain an approximate value for Ne' from those readings alone. It was therefore not necessary to compare only grains of the same size and shape.

These various facts considerably enlarge the notion of elementary charge introduced by Helmholtz. Moreover, whereas electrolysis has not up to the present suggested any means for measuring directly the absolute value of the charge e on a monovalent ion, we shall see that it is possible to measure that charge when it is carried by a microscopic

---

[1] Comptes Rendus, Vol. CXLVI., 1908, and Le Radium, 1909.
[2] Physik. Zeitschrift, Vol. XII., 1911, p. 630.

granule in a gas. In this way we shall obtain, since Ne is known, a fresh determination of N and of the molecular magnitudes.

99.—DIRECT DETERMINATION OF THE IONIC CHARGE IN GASES.—If an ion in a gas is brought by the molecular agitation into the vicinity of a speck of dust, it will be attracted by induction and will attach itself to the speck, charging it in consequence. The arrival of a second ion of the same sign will be checked by the repulsion between the two charges, and will also be the less likely to occur the smaller the speck of dust.[1] The arrival of an ion of opposite sign will, on the contrary, be facilitated. A number of the dust particles will therefore either remain neutral or will become neutral again subsequently, and a permanent equilibrium will be set up if the ionising radiation continues to act. This has actually been demonstrated to be the case for various kinds of smoke particles, neutral to begin with, when the gas wherein they are suspended is ionised (de Broglie).

Another interesting case is that of an ionised gas, free from dust particles but saturated with water vapour. C. T. R. Wilson's experiments (1897) prove that the ions act as centres of condensation for the water droplets that make up the mist that forms when the gas is cooled by an adiabatic expansion.

Finally, a gas can be charged simply by bubbling through a liquid (which involves the rupture of liquid membranes). The formation of charged mists in gases prepared by electrolysis, first noticed by Townsend, is probably caused in this way.

In any one of the above cases the elementary charge e will be determined if the charge acquired by the drop or dust speck can be measured. The first determinations of this charge were made by Townsend and J. J. Thomson (1898). Townsend worked on the mists carried along by gases produced in electrolysis, while Thomson used the

[1] More strictly it will rarely happen that the molecular agitation will impart a sufficient velocity to an ion to enable it to penetrate into the region where the attraction of the speck due to induction will overcome the repulsion. The theory of electric images enables a definite calculation to be made.

clouds formed on the condensation of ionised damp air by expansion. They determined the total charge e present in the form of ions in the cloud under investigation, the weight P of the cloud, and finally its rate of fall v. This latter measurement gave the radius of the drops (assuming that Stokes' law is applicable to them) and hence the weight p of each. Dividing P by p we get the number of drops n and hence the number of ions. Finally, the quotient of E by n gives the charge e. The number obtained in Townsend's experiments, which obviously were not very exact, varied between 1×10^{-10} and 3×10^{-10}; Thomson's varied between $6 \cdot 8 \times 10^{-10}$ (using the negative ions emitted by zinc illuminated by ultra-violet light) and $3 \cdot 4 \times 10^{-10}$ (with the ions produced in a gas by X rays or the rays from radium). These values approximated well enough as to the order of magnitude of the expected result, and, although the agreement was still rather rough, it was of great importance at the time.

The method, employed in this way, involved a high degree of uncertainty. It was assumed, for instance, that every ion is united with a droplet and that each droplet carries only one ion.

H. A. Wilson simplified the method very considerably (1903). He confined himself to the measurement of the rates of fall of the cloud, first when gravity operates alone, and then when it is opposed by an electrostatic force. Let v and v' be the velocities, before and after the application of an electric field H, of a droplet bearing a charge e' and weighing mg. Making the single hypothesis that these constant velocities are proportional to the operative forces, we get (H. A. Wilson's equation) *even if Stokes' law is inexact* :

$$\frac{\text{H}e' - mg}{mg} = \frac{v'}{v},$$

or
$$e' = m \cdot \frac{g}{\text{H}} \cdot \left(\frac{v + v'}{v}\right).$$

Further, during the uniform fall of a drop, the motive force (its weight $\frac{4}{3}\pi a^3 g$) is equal to the frictional force and hence to

THE ATOM OF ELECTRICITY 181

$6\pi a \xi v$ *if Stokes' law is valid.* This gives the radius and hence the mass m, so that the charge e' can be calculated.

Under the influence of the field, the charged cloud obtained by the expansion of air (strongly ionised) divides itself into two or even three clouds sinking at different rates. Applying the above equations to the motions of these clouds (regarded as being composed of identical droplets), values roughly proportional to 1, 2, and 3 were obtained for the charge e'. This proves the existence of *polyvalent* drops. The value found for the charge e' for the least charged cloud varied between $2 \cdot 7 \times 10^{-10}$ and $4 \cdot 4 \times 10^{-10}$, the mean value being $3 \cdot 1 \times 10^{-10}$. The want of precision is thus still great. Fresh experiments were carried out, using the same form of apparatus, by Przibram [alcohol droplets], who found $3 \cdot 8 \times 10^{-10}$; he was followed by other physicists. The latest and most trustworthy result (Begeman, 1910) gives $4 \cdot 6 \times 10^{-10}$ (Stokes' law being always assumed). We shall see that the measurements are very greatly facilitated by studying the charged particles *individually*.

100.—The Study of the Individual Charges Proves the Atomic Structure of Electricity.—H. A. Wilson's reasoning refers to a single particle. Now, in the experiments described above, it is applied to a cloud, and it is assumed that the droplets in the cloud are identical, which is certainly incorrect. All uncertainty of this kind is avoided by working under the experimental conditions postulated in the theoretical treatment of the question ; in other words, by observing a single spherule, infinitely removed from all other spherules and from the walls of any enclosure.

Observations on individual charged grains, thus correctly applying the method invented by H. A. Wilson, were made independently (1909) by Millikan and by Ehrenhaft.

Ehrenhaft, however, working with dust particles (obtained by sparking between metals), unfortunately applied Stokes' law, regarding his particles, without proof, as complete homogeneous spheres. I am inclined to think that they are really irregular, spongy bodies having an entirely irregular and jagged surface ; their frictional effect in gases will be very much greater than if they were spheres, and the application of Stokes' law to them has no meaning. I regard as proof of this the fact, pointed out by Ehrenhaft himself, that many of these dust clouds have no appreciable Brownian movement, although they are ultra-microscopic. This observation, which has received little attention, indicates an enormous frictional effect. And, in fact, Weiss's recent measurements referred to above (para. 98) show that dust particles that, according to Ehrenhaft, should bear very small charges lying between 1×10^{-10} and 2×10^{-10}, show displacements which give quite normal values for Ne. These dust particles therefore carry charges in the neighbourhood of $4 \cdot 5 \times 10^{-10}$.

Millikan, working with droplets that certainly possessed a massive, close-grained structure (obtained by atomising liquids), has carried out experiments that are free from the objections referred to above. The droplets are carried by a current of air to the neighbourhood of a pin-hole pierced in the upper plate of a flat horizontal condenser. A few of them pass through the hole and, when between the condenser plates, are illuminated laterally and can be followed by means of an eye-piece (as in de Broglie's apparatus); they then appear as brilliant points of light on a black background. The electrostatic field, of the order of 4,000 volts per centimetre, acts counter to gravity and generally prevails over it. It is then possible to balance the same droplet for *several hours* without losing sight of it, alternately making it rise under the influence of the field and letting it sink by cutting it off.[1] Since the droplet, being composed of a non-volatile substance, remains the same throughout, its rate of fall has always the same constant value v. Similarly its upward motion takes place at a constant velocity v'. But in the course of a long series of observations, it sometimes happens that the upward velocity *suddenly* changes, *in a discontinuous manner*, from the value v' to another value v_1', which may be greater or less than v'. The charge on the droplet has therefore changed, *in a discontinuous manner*, from e' to another value e_1'. This discontinuous variation becomes more frequent when the gas in which the droplet

[1] For full details of Millikan's work, see Physical Review, 1911, pp. 349–397.

THE ATOM OF ELECTRICITY

is moving is subjected to an ionising radiation. It therefore seems natural to attribute the variation of the charge to the fact that an ion, when near the dust particle, gets captured by electric attraction, in the way explained above.

Millikan's remarkable observations demonstrate in a vigorous and direct manner the atomic structure assumed for electricity. Writing down H. A. Wilson's equation for the condition of affairs before and after the discontinuous change and dividing the one by the other, we get

$$\frac{e'}{e_1'} = \frac{v+v'}{v+v_1'},$$

or, better,

$$\frac{e'}{v+v'} = \frac{e_1'}{v+v_1'} = \frac{e_2'}{v+v_2'} = \ldots \ldots$$

for the ratio between the charges e' and e_1'. The successive charges borne by the drop must therefore be whole-number multiples of the same elementary charge e, if the sums $(v + v')$, $(v + v_1')$, etc. are proportional to whole numbers (if, that is to say, they are equal to the products obtained by multiplying various whole numbers by the same factor). Moreover, the whole numbers corresponding to two successive charges will differ in general by one unit only, corresponding to the addition of one elementary charge (it being nevertheless possible for a polyvalent ion to be formed).

These conclusions can be checked by means of Millikan's figures[1]. For instance, the successive values of $(v + v')$, and consequently of the successive charges, for a certain oil drop, were to one another as the following numbers:

2·00, 4·01, 3·01, 2·00, 1·00, 1·99, 2·98, 1·00 :

that is to say, they are to one another, to within 1 per cent., as

2, 4, 3, 2, 1, 2, 3, 1.

---

[1] As a matter of fact Millikan presents his results in a different form, giving at once the absolute values of the charges obtained by combining Stokes' law with H. A. Wilson's equation. In my opinion, it is of greater advantage to put forward first of all the facts that would be unassailable even if Stokes' law were quite inapplicable to droplets falling in a gas.

For another drop, the successive charges, as indicated by the velocities, were again proportional to whole numbers :

5, 6, 7, 8, 7, 6, 5, 4, 5, 6, 5, 4, 6, 5, 4,

with a variation of the order of 1 in 300, which is the limit of precision set by the accuracy of the measurements of velocity.

As Millikan points out, such a degree of precision is comparable with that which satisfies chemists in verification of the laws of discontinuity resulting from the atomic structure of matter.

The numerical examples just quoted show that the moments when a given drop is carrying a single elementary charge can be very quickly recognised. If at such a time the activity $\dfrac{X^2}{t}$ of its Brownian movement is measured by de Broglie's or Weiss's methods (para. 98), the product Ne can then be derived from Townsend's equation. This has been done by Fletcher in Millikan's laboratory ; 1,700 determinations, divided among 9 drops, give $28 \cdot 8 \times 10^{13}$ for the product, which agrees to within 1 part in 200 with the value given by electrolysis.

In short, Millikan's experiments demonstrate in a decisive manner the existence of an atom of electricity equal to the charge carried by the hydrogen ion during electrolysis.

101.—THE VALUE OF THE ELEMENTARY CHARGE : DISCUSSION.—The precise value of this elementary charge, which we now know to exist, has yet to be determined. To do this the mass m of the droplet must be measured, since H. A. Wilson's equation gives the ratio $\dfrac{e}{m}$ only, and up to the present no better means has been found than the application of Stokes' law

$$mg = 6\pi a \xi v$$

with the introduction of suitable corrections.

There is no doubt, in the first place, that the product $6\pi a \xi v$ does not accurately express the frictional force acting upon a microscopic spherule moving in a gas at a velocity v. The expression holds for liquids (para. 61), but in that case the radius is very great in comparison with the mean free path of the fluid molecules; whereas in gases it is of the same order of magnitude. The frictional effect will in consequence be lessened, as will be apparent when we consider that if L were to become very great, or if, which comes to the same thing, the gas were to disappear, there would be no friction at all, although the formula indicates that the friction is independent of the pressure.[1] A very complete theory, put forward by Cunningham, leads us to take, for the value of the frictional force, the product $6\pi a \xi v$ divided by

$$1 + 1\cdot 63 \frac{L}{a} \cdot \frac{1}{2-f},$$

f being the ratio of the number of molecular impacts that are followed by regular reflection (elastic impacts) to the total number of impacts undergone by the spherule.

Millikan made the single assumption that the frictional force must be of the form

$$\frac{6\pi a \xi v}{1 + \alpha \frac{L}{a}},$$

and worked out the value of the constant α that would give approximately equal values of e for each drop. With α equal to $\cdot 81$,[2] the values of $e \times 10^{10}$ for different drops fall between 4·86 and 4·92 (the uncorrected values lying between 4·7 and 7). Millikan therefore concluded that the value of e is $4\cdot 89 \times 10^{-10}$, which gives the value

$$59 \times 10^{22}$$

for N, which is, after all, in remarkable agreement with the value 68×10^{22} given above.

---

[1] Compare para. 48, note 2.
[2] This value follows from Cunningham's equation for the case where f is zero (*i.e.*, for a perfectly rough sphere). This assumption of perfect roughness seems to me to present difficulties. A shot fired obliquely against a surface itself made up of shot packed together almost perfectly might rebound back along its line of incidence, but such a case would be exceptional. Moreover, the mean direction of reflection even if it were not the direction of regular reflection, cannot be far short of it.

Millikan regards the error in his result as being less than 1 in 2,000, though in my opinion so high a degree of precision is very doubtful,[1] on account of the magnitude of the correction that has to be applied to Stokes' law before we can deduce the mass of a spherule from its rate of fall in air.

M. Roux undertook to repeat the experiments in my laboratory, and to measure, at my suggestion, the rate of fall of *the same spherule* in air and in a liquid.[2] Since Stokes' law applied in the case of a liquid, measurements in the latter case give, *without correction*, the exact radius of the spherule.

M. Roux made use principally of spherules of super-cooled, pulverised sulphur, which were glass-like at the ordinary temperature, and found that Cunningham's formula is applicable, though with the coefficient f nearly equal to 1 (the surface of the spherules being therefore polished rather than rough). He then, like Millikan, followed the same spherule for several hours under the microscope, as it sank under the influence of gravity, rose under the influence of the electric field, and from time to time suddenly gained or lost an electron.

In this way he found that the value of the charge e lay between 4×10^{-10} and $4 \cdot 4 \times 10^{-10}$, or, using the equivalent form, N had the value 69×10^{22} to within ± 5 per cent. This result is practically identical with that obtained from my study of the Brownian movement. Applying to Millikan's rough results the corrections which M. Roux's experiments have made legitimate, we get the value 65×10^{22} for N. I shall adopt 67×10^{22} as the mean value given by this method.

102.—CORPUSCLES : SIR J. J. THOMSON'S RESEARCHES.— Sir J. J. Thomson's brilliant work has shown that the atom of electricity, the existence of which has just been established, is an essential constituent of matter.

The electrical nature of the kathode rays once established, attention was turned to the solution of the two equations

---

[1] More especially since Millikan, in a later publication, has himself raised the value he proposed for N by 2 per cent.

[2] For details of these difficult experiments see Roux, Ann. de Chim. et Phys., 1913.

THE ATOM OF ELECTRICITY 187

which, by the application of the known laws of electrodynamics, can be shown to express the electrostatic and magnetic deviations undergone by an electrified projectile bearing a charge e, of mass m, and moving with velocity v; these deviations remain constant so long as the ratio $\frac{e}{m}$ keeps a constant value.

In this way Thomson found values for the velocity of the order of 50,000 kilometres per second in an ordinary Crookes' tube; they depended, moreover, very much on the potential difference used to produce the discharge (in the same way that the velocity of a falling stone depends on the height of its fall).

But the ratio $\frac{e}{m}$ is independent of all circumstances whatever, and according to the best measurements [1] is 1,830 times greater than the ratio of the charge to the mass for the hydrogen ion during electrolysis. The same value is found whatever the nature of the gas through which the discharge passes and whatever the nature of the electrode metals; it is the same for Lenard's slow kathode rays (1,000 kilometres per second) that, without any discharge, are emitted by metallic surfaces (zinc, the alkali metals, etc.) when acted on by ultra-violet light. In explanation of this universal result Thomson suggested (and all subsequent facts have tended to confirm his theory) that the kathode projectiles are always identical and that each of them carries a single atom of negative electricity; each of them is consequently about 1,800 times lighter than the lightest of all atoms. Moreover, since they can be produced from any kind of matter, that is to say, from any kind of atom, these material elements must be a universal constituent common to all atoms; Thomson proposed to call them *corpuscles*.

A corpuscle cannot be considered independently of the negative charge it carries; it is inseparable from that charge and the charge *constitutes* the corpuscle.

Incidentally, the high conductivity of metals is very

[1] Classen, Cotton and Weiss, etc. I have allowed for the fact that the faraday is equal to 96,600 coulombs (and not to 100,000).

simply explained (Thomson, Drude) if it is assumed that at least a few of the corpuscles present in their atoms can be displaced by the action of even the feeblest electric fields, passing from one atom to another or even moving hither and thither in the metallic mass as freely as the molecules in a gas.[1]. When we remember to what extent matter is really empty and hollow (para. 95), this hypothesis will not greatly perturb us. The electric current, which in electrolytes consists in the movement of charged atoms, becomes in metals a stream of corpuscles, that can produce no chemical effect on passing through a zinc-copper junction, since corpuscles are the same in zinc and copper. The action of a magnet on a current does not differ fundamentally from the action of a magnet on kathode rays. Laplace's law, moreover, at once gives, at all points, the value and direction of the magnetic effect produced by a single moving charged particle,[2]

As long as we were unable to measure even approximately the charge on a single projectile, and as long as we hesitated to accept the proof of the view that such projectiles are fragments of atoms, there were certainly grounds for the objection that the high value found for $\frac{e}{m}$ might be equally well explained as being due to the magnitude of the charge as to the smallness of the mass. But, as we have seen (para. 99), Thomson obtained accurate measurements of the charge on the droplets obtained by the expansion of damp air containing no other ions than the negative corpuscles detached from a metallic surface by ultra-violet light. If the charge borne by these corpuscles is equivalent to 1,800 electrons, the charge on a droplet should be *at least* 1,800 times greater than the charge actually found.[3]

[1] A more detailed analysis shows that the other essential properties of the metallic state can be explained in the same way (opacity, metallic lustre, thermal conductivity).

[2] Laplace's classic expression $\frac{Ids \sin\alpha}{r^2}$ gives the value $\frac{e.v.\sin\alpha}{r^2}$ for the field due to a *single* particle.

[3] It is easy to prove that the droplets in a cloud have captured all the electric charges in the gas and, by the application of an electric field, to ascertain that the droplets themselves are all charged; those among them that are polyvalent (H. A. Wilson) would therefore be carrying several times 1,800 electrons.

THE ATOM OF ELECTRICITY

We are thus forced to the conclusion that the corpuscle has a mass very much less than that of any known atom. To be precise, the mass of this natural element, the smallest we have hitherto discovered, is the quotient of the mass of the hydrogen atom by 1,835, which in grammes comes to

$$8 \times 10^{-28}.$$

Thomson has gone a step further and has finally been able to give us some idea of the *corpuscular dimensions*. The kinetic energy $\frac{1}{2}mv^2$ of a moving corpuscle must exceed the magnetic energy produced in space as a result of its motion. In this way [1] it is found that the corpuscular diameter must be less than $\frac{1}{3} \times 10^{-12}$; less, that is to say, than the hundred-thousandth part of the diameter of impact of the smallest atoms.

For reasons that I cannot enter into here, it is probable that the upper limit to the magnetic energy is actually reached; in other words, the *whole of the inertia* of the corpuscle is due to the magnetic effect that accompanies its motion like a wake. It is possible that this may be the case for all matter, even when neutral, if its neutrality is simply the result of equivalence between charges of opposite sign to be found therein (such charges may even be the sole constituents of matter). All inertia would then be electromagnetic in its origin. I cannot explain here how the mass of a projectile, which remains sensibly constant so long as its speed does not exceed 100,000 kilometres per second, would above that speed actually increase as the speed increases, slowly at first and then more and more rapidly, finally becoming infinite at a speed equal to that of light; which means that no matter can reach that velocity (H. A. Lorentz).

103.—POSITIVE RAYS.—Besides the kathode rays and

---

[1] We have only to integrate the expression $\frac{H^2.dv}{8\pi}$ over all space outside a corpuscle of radius a, remembering that (Laplace's law) the magnetic field H is equal to $\frac{ev \sin \alpha}{r^2}$ at any point.

X rays, a third variety, of equal importance, has been observed in Crookes' tubes.

Before the pressure becomes very low (one-tenth of a millimetre) a luminous *sheath* (violet in air) surrounds the kathode but does not touch it (like an equipotential surface). It moves from the kathode and shades away from it as the pressure falls ; but its interior contour remains fairly sharp and is found 1 or 2 centimetres away from the kathode when the kathode rays have become vigorous.

A second fairly bright luminosity (orange in air) now

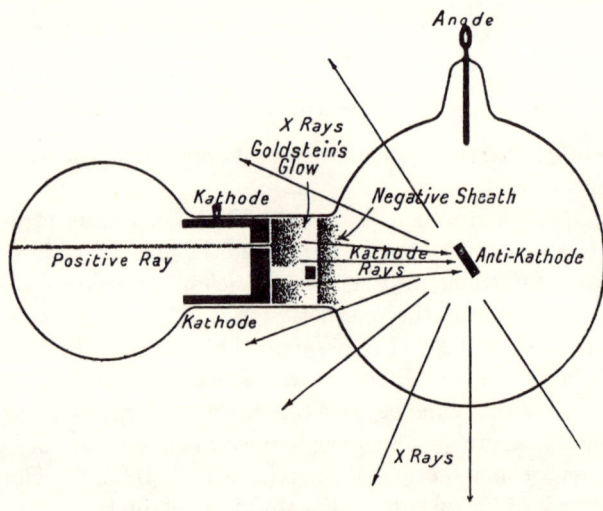

Fig. 13.

becomes visible close up to the surface of the kathode ; it extends, gradually becoming paler, several millimetres from the kathode. It is produced by *rays* coming from the inside contour of the sheath (for an obstacle inside the contour throws a shadow upon the kathode, no shadow being formed when the obstacle is without the sheath). The velocity of these rays probably increases as they pass from the sheath to the kathode (so that a maximum luminosity is produced near it).

In order to observe these rays better, Goldstein conceived the happy idea (1886) of piercing a *canal* through the kathode

THE ATOM OF ELECTRICITY

on which they strike. Provided that the kathode divides the tube into two parts (it has since been found that the essential condition is that the space lying behind the kathode should be protected from electrical effects), a ray passing through the canal is able to travel for several centimetres into the second part of the tube, and finally to indicate the point where it arrives on the glass wall by a pale fluorescence.

I have attempted to represent in the diagram the relations between the three kinds of rays produced in Crookes' tubes.[1]

The electrical nature of the rays once recognised, it was natural to inquire whether the rays discovered by Goldstein, which strike against the kathode instead of starting from it, are not positively charged.

That this is the case, as seemed very probable from certain observations of Villard's, was demonstrated by Wien. He established the fact that the rays are deviated by an electric field as though they were a stream of electricity (on being passed between the two surfaces of a small condenser); they are also deflected by a magnetic field (which is very much less effective than with kathode rays). Measurements of the velocity and of the ratio $\frac{e}{m}$ are therefore possible.

The velocity, which varies according to the conditions, is only a few hundred kilometres per second. The ratio of charge to mass turns out to be of the same order of magnitude as in electrolysis; the positive rays are therefore composed of ordinary atoms (or groups of atoms).

The measurements mentioned above were rough, for the positive rays become very indistinct when deviated. This fact may be accounted for (Thomson) on the supposition that an atom projected at very high speed may, when it strikes a neutral molecule, lose (or gain) fresh corpuscles.[2] If this happens while the projectile is passing through the deviating field, the deviation may become anything. By

---

[1] In addition, kathode rays start only from these points on the kathode that are struck by Goldstein rays (Villard).
[2] The fact that the positive rays render the rarefied gases through which they pass conducting, by leaving ionised molecules in their track, is explained at the same time.

hermetically sealing the kathode into the wall of the tube, Thomson arranged matters so that communication between an observation chamber and the emission chamber was maintained by means of the canal through which passed the pencil of rays under examination. This canal was so long and so narrow that it was possible to *maintain* a much higher vacuum in the observation chamber than in the emission chamber. Encounters during the passage through the field are thus practically eliminated and, if the (common) direction of the electric and magnetic fields is perpendicular to the pencil of rays, projectiles of the same kind (*i.e.*, those having the same $\frac{e}{m}$ but *moving with different velocities*) must strike a plate placed opposite the canal at various points on the same parabola. Conversely, each parabola that appears on the plate determines (to within 1 per cent.) the ratio $\frac{e}{m}$ for a particular kind of projectile.

The mass m, multiplied by a factor (which in general can readily be determined), is thus known, and in this way molecules of an unexpected type have been shown to exist ("a new chemistry," in the words of J. J. Thomson). Remarkable research in this field has been carried out by Aston, who, by producing positive rays in chlorine, has found atomic projectiles of masses 35 and 37, and none of mass 35·5. In the same way Aston has been able to separate into isotopes the elements lithium, boron, magnesium, silicon, potassium, cobalt, bromine, mercury, neon, argon, krypton and xenon, thus verifying Prout's hypothesis.

Positive electrons lighter than the hydrogen atom have never been isolated; the various kinds of ionisation always divide the atom into one or more negative corpuscles of insignificant mass on the one hand, and a positive ion, relatively very heavy and containing the rest of the atom, on the other.

The atom is therefore not indivisible, in the strict sense of the word, but consists possibly of a kind of positive sun, or nucleus, bearing a fixed charge wherein resides its *chemical individuality*, and about which swarms a cloud of negative planets, of the same kind for all atoms.[1] It has been possible to calculate the number of these "planets" from the lateral diffusion (measurable) they cause in the X rays falling on a known number of atoms (Barkla). The very remarkable result is obtained that the number of negative electrons is equal to the atomic number (sequence number) P of the atom (para. 17). The hydrogen atom should thus be composed of a positive nucleus (Prout's proto-atom), with a single negative electron circulating round it. In the case of two isotopic atoms the nuclei have different masses, but the same charges, and are surrounded by *identical* planetary swarms.

104.—MAGNETONS.—This rough model suggests that revolving corpuscles, which would be equivalent to circular currents, probably exist within the atoms. Now, a circular current (solenoid) possesses the properties of a magnet. We are reminded of the hypothesis that explains magnetism by supposing that the molecules of a magnetic body are small magnets (Weber and Ampère).

Langevin regards thermal agitation as the cause that prevents these little magnets from arranging themselves parallel to each other in the feeblest magnetic field, in which case the body would at once reach its maximum magnetisation (its moment M_o per gramme molecule being equivalent to N times the amount of a single molecule). By assuming that statistical equilibrium [2] is reached between the de-orientation caused by thermal agitation and the orientation caused by the field H, Langevin has been able to calculate the maximum magnetisation M_o from the observed magnetisation, of moment M, produced by that field [3] (1005).

This brilliant theory was put forward for the case of a feebly magnetic fluid. Pierre Weiss completed it and showed

---

[1] As Rutherford has been kind enough to point out, this model was first suggested by myself ("Les Hypothèses Moléculaires," *Revue Scientifique*, 1901), but without serious proof.

[2] An analogous theory explains electrical double refraction (Kerr) and the magnetic double refraction recently discovered by Cotton and Mouton.

[3] The observed moment M is, for low values of $\frac{H}{RT}$, equal to $\frac{M_o.H}{3RT}$ (the law expressing the influence of temperature, discovered previously by Curie, may here be recognised).

that his results hold good even for solids. He moreover explained ferro-magnetism (by the hypothesis of a very intense internal field, caused by mutual action between molecules) Finally, he was able to deduce from his experiments the values of the maximum moment M_o per gramme molecule for various atoms. In so doing he made the very important discovery that these values are whole number multiples of the same number (1,123 C.G.S. units).

Weiss found himself forced to the conclusion that in all paramagnetic atoms small identical magnets exist. These magnets he called *magnetons*.[1] These magnetons may be arranged in file or in parallel positions, in which case their moments would be added together, although they might also oppose each other in astatic couples having no external effect.

These magnets, however, would lie in the periphery of the atom; in fact, measurements of magnetisation show that small chemical and physical changes are able to alter the number of atomic magnetons.

[1] Investigations in the magnetic field have been carried out for the atoms of Fe, Ni, Co, Cr, Mn, V, Cu.

CHAPTER VIII

THE GENESIS AND DESTRUCTION OF ATOMS

TRANSMUTATIONS.

105.—RADIOACTIVITY.—The study of the electric discharge through rarefied gases has led to the discovery of three kinds of radiations, all of which possess the common characteristics of affecting photographic plates, of exciting various kinds of fluorescence, and of making conductors of the gases through which they pass.

Certain substances are known that, without external excitation, continually emit rays analogous to those obtained in rarefied gases. This most important discovery was made in 1896 by Henri Becquerel with respect to uranium compounds and to metallic uranium itself. The *uranium* rays have a feeble though constant intensity, which is the same in light or in darkness, at high or low temperatures, at midday or at midnight.[1] Their intensity depends only on the mass of uranium present, and not in the least on its state of combination. Two different uraniferous substances, spread out in *very* thin layers (to prevent absorption in the layer) in such a way that the same quantity of uranium is present per square centimetre, will emit the same amount of radiation from equal surfaces. We are therefore dealing with an *atomic property*; wherever uranium atoms are to be found, there energy is being constantly emitted. In this we find the first indication that something may be happening in the interior of the atom and *that the atoms themselves are not immutable* (Pierre and Marie Curie). To this atomic property the name of *Radioactivity* has been given.[2]

---

[1] This fact, which was established by Curie, eliminates the hypothesis of an excited radiation caused by an invisible solar radiation.
[2] This word was introduced by Mme. Curie. Of course, a substance is not radioactive (any more than a Crookes' tube is) when it emits ionising rays

It appeared hardly probable that this property would be found to be associated with uranium alone. On all sides systematic examinations of all the known elements were undertaken. Schmidt was the first to record the radioactivity of thorium and its compounds, which have an activity comparable with that of uranium. More recently, as a result of the high state of perfection reached in the methods of measurement, it has been possible to demonstrate radioactivity, definite though about 1,000 times more feeble, in the case of potassium and rubidium. It is permissible to suppose that all atoms are radioactive, though in very different degrees.

It occurred to Mme. Curie to examine the natural minerals, as well as the pure substances obtained from them. She then noticed that certain rocks (notably pitchblende) are about 8 times more active than their uranium and thorium content would lead one to suppose, and came to the conclusion that this fact pointed to the presence of traces of strongly radioactive unknown elements. It is well known how this hypothesis was verified and how, by solution and fractional crystallisation (the successive stages in the purification being followed with the electrometer), Pierre Curie and Mme. Curie obtained, starting from various uraniferous minerals, products that became, as their purification progressed, more and more radioactive, then luminous by auto-fluorescence, and finally yielded pure salts of a new alkaline earth metal, *radium*. The atomic weight of the new metal is equal to 226·5, and it is analogous to barium in its spectrum and in its general properties (radioactivity excepted). It is at least a million times more radioactive than uranium (1898—1902). In the course of this work Mme. Curie had detected, without isolating it, another strongly radioactive element, *polonium*, analogous to bismuth, and shortly afterwards Debierne established the existence in the same minerals of an element *actinium* that accompanies the rare earths in the fractionations.

With the very active preparations that could now be

in a purely temporary manner, as the result, for instance, of a chemical reaction (glowing metals, phosphorus in process of oxidation, etc.).

GENESIS AND DESTRUCTION OF ATOMS

obtained, the radiation could easily be analysed; the three types of radiation discovered in Crookes' tubes were found at once and could be studied by similar methods. The three types are:

α rays or positive rays (Rutherford) composed of positively charged projectiles, having masses of the same order of magnitude as the various atomic masses. Their velocity may exceed 20,000 kilometres per second and they are consequently much more penetrating than Goldstein's rays; they are nevertheless stopped completely after travelling a few centimetres in air;

β rays or negative rays (Giesel, Meyer and Schweidler, Becquerel), composed of corpuscles moving at speeds that may exceed nine-tenths that of light. These rays are very penetrating kathode rays and lose scarcely half their intensity after a path in air of the order of a metre;

γ rays, which cannot be deviated (Villard) and are extremely penetrating, passing through 1 centimetre of lead before their intensity is halved. They are very analogous to X rays, and undoubtedly do not differ from them more than blue light differs from red.

These three radiations, each having properties varying according to the nature of the radioactive source, are not emitted in a constant ratio to each other, and generally speaking, are not even all emitted by the same element (for instance, polonium emits practically only α rays).

Pierre Curie discovered (1903) that the total energy radiated, which is measurable in a calorimeter having absorbing walls, has an enormous value, and is independent of the temperature. A closed tube containing radium liberates, when in radioactive equilibrium, 130 calories per hour per gramme of radium. Expressed differently, it liberates in about two days, without appreciable change, as much heat as would be produced by the combustion of an equal weight of carbon. It has thus become possible to trace the origin of the internal heat of the earth and of the radiation of the sun and stars to radioactive sources.

106.—RADIOACTIVITY AS THE MANIFESTATION OF ATOMIC DISINTEGRATION.—Pierre and Mme. Curie noticed (1899) that solid objects placed in the same enclosure as a salt containing radium (arranged in such a way that the path from the salt to the object lay entirely through the air) appeared to become radioactive also; this *induced radioactivity*, which was independent of the nature of the object, gradually decays when the object is withdrawn from the influence of the radium and becomes practically nothing after a day. Rutherford soon after observed the same property in connection with thorium, which excites an induced radioactivity of slightly longer duration.

These induced radioactivities are produced at all points to which a gas liberated by the original radioactive preparation might penetrate by diffusion. It occurred to Rutherford that material gaseous *emanations* might actually be continuously engendered by radium and thorium. On aspirating off the air that had remained in contact with a thorium salt, he found that the air remained a conductor, as though it were preserving some internal source of ionisation. This spontaneous ionisation decreases in geometrical progression, being reduced to about half its value every minute. The same effect is noticed in air that has passed over a radium salt, though the rate of decay is slower, diminution to about half value occurring after four-day intervals.

Rutherford then assumed that the radioactivity of an element does not indicate the *presence* of atoms of the element, but their *disappearance* or *transformation* into atoms of another kind. The radioactivity of radium, for instance, implies the destruction of radium atoms and the appearance of atoms of emanation; and although a given mass of radium seems to us to be invariable, this is so only because our measurements do not extend over a sufficiently long period of time. The radioactivity of the emanation implies the destruction of atoms of that gas, at the rate 1 out of 2 in four days, new atoms appearing that form a *solid* deposit on objects that come in contact with the emanation. The atoms of the deposit die in their turn, at the rate of 1 out of 2 in about half an hour, which explains the induced radioactivity mentioned above; and so on.

Rutherford's ingenious suggestions have been completely

GENESIS AND DESTRUCTION OF ATOMS

verified. A *radium emanation* can be isolated and is continuously liberated by radium at the rate of a tenth of a cubic millimetre per day per gramme. This gas liquefies at $-65°$ C. and solidifies at $-71°$ C. (to a self-luminous solid). It is chemically inert, like argon, and hence *monatomic* (Rutherford and Soddy); its density (Ramsay and Gray) and its rate of effusion through a small hole (Debierne) indicate in that case an atomic weight of about 222. When caused to glow by an electric discharge it gives a characteristic line spectrum (Rutherford). In short, it is a definite chemical element, which Ramsay has proposed to call *Niton* (shining). It is, however, an element that decays spontaneously to the extent of one-half in four-day intervals (more exactly, after intervals of 3·85 days). For the first time the fact has been established that a simple substance, and hence an atom, can be born and can die.

We can now scarcely avoid the conclusion that radium itself also gradually decays, and at the same rate that it produces niton, or very nearly at the rate of one-thousandth of a milligramme per day per gramme. In short, we are led to the conclusion that all radioactivity is the sign of the *transmutation* of an atom into one or more other atoms.

These transmutations are *discontinuous*. We find, for instance, no intermediate steps between radium and niton; we either have radium atoms or niton atoms, and we can obtain no evidence of any matter that has ceased to be radium and has not become niton. Similarly, for as long as it is possible to observe niton, that gas retains its particular properties, whatever its "age," and continues to disappear to the extent of one-half after every four-day interval. Transmutation must occur atom by atom, suddenly and *explosively*, and it is during the explosions that the various rays are shot out. When, for instance, we say that the radioactivity of uranium is an atomic property, we do not mean that all the uranium atoms present are concerned, but only those that are actually disintegrating (the number of the latter being moreover proportional at each instant to the mass of uranium present). It is only at the very moment of explosion that an atom is radioactive.

107.—The Production of Helium.—We should not perhaps have been so ready to accept Rutherford's views were it not that a certain simple substance already known was found to be produced by transmutation. Ramsay and Soddy succeeded in proving that *helium* is produced in ever-increasing quantity in a closed vessel containing radium (as, indeed, Rutherford and Soddy had predicted would be the case). This brilliant piece of experimental work removed all doubt in the minds of physicists as to the possibility of spontaneous transmutation (1903).

It was known, moreover, that the α particles have masses of the same order as the atomic masses. More precisely, the ratio $\frac{e}{m}$ is always very nearly the same from whatever element the rays are produced, and is equal to about half the value of the same ratio for the hydrogen ion in electrolysis. The α particles might therefore be atoms having a coefficient equal to 2; but they might also (Rutherford) be helium atoms carrying two elementary charges each. That the second alternative is correct was proved directly by Rutherford and Roys. They enclosed some niton in a thin-walled glass vessel (the thickness of the walls was of the order of a hundredth of a millimetre), through which molecules of a gas possessing the degree of agitation corresponding to the ordinary temperature were unable to pass (this had been demonstrated for helium in particular), though the α rays emitted by niton could pass through easily. Under these conditions helium is soon found in the outer vessel into which the rays have penetrated; *the α projectiles are helium atoms expelled at the prodigious speed of* 20,000 *kilometres per second*.

108.—α Rays.—The atomic weight of radium is very nearly the sum of the atomic weights of niton and helium. During transmutation, therefore, the radium atom splits up into a helium atom and a niton atom, with an explosion that expels the helium atom to a distance, and which must also propel the niton atom in the opposite direction, with an equal quantity of motion (a phenomenon analogous to the recoil of a gun). The initial velocity of this niton projectile

GENESIS AND DESTRUCTION OF ATOMS

can therefore be easily calculated, and is found to be several hundred kilometres per second.

I do not see that there is any reason for drawing a distinction between the two projectiles; slow α rays composed of niton (very similar to Goldstein's rays) must be taken into account as well as α rays of helium. I shall return to this point later.

109.—A TRANSMUTATION IS NOT A CHEMICAL REACTION.— A first account of the splitting up of radium into helium and niton raises the question whether that change may be regarded as a chemical reaction that certainly disengages much heat, but which is nevertheless not essentially different from ordinary reactions. Why cannot radium be considered as a compound yielding niton and helium on dissociation ?

This attitude cannot be maintained when it becomes apparent that all the factors that influence chemical reaction are found to be of no effect in the case of radioactive change. A rise in temperature of 10° is just about enough to double the speed of a reaction. At this rate a reaction should become 10,000,000,000 times faster for each elevation of 300° C. Now, the heat liberated by radium remains absolutely unaffected by much greater temperature variations.

This behaviour is general. By no means whatever is it possible to modify the inflexible course of radioactive transformation. Heat, light, magnetic field, high concentration, or extreme dilution of the radioactive material (that is to say, intense or negligible bombardment by α and β projectiles) have no effect. Deep within the atom, in the highly condensed *nucleus* which has been shown to exist therein, a disintegration takes place that is affected as little by influences we can control as is the evolution of a distant star. We may add that the explosions of two atoms of the same kind appear to be absolutely identical, giving exactly the same velocities to the emitted α projectiles (and also to the β projectiles).

110.—ATOMS DO NOT DECAY.—We can go even further and catch a glimpse of the infinitely complex world within the nucleus.

We have seen that whatever the age of a given mass of niton, half of that mass disappears in four days. The atoms therefore do not decay, since every atom that escapes destruction (during any given time) still has an even chance of survival for the four days following.

Similarly, if two small globes connected by a tube were to contain a mixture of oxygen and nitrogen in statistical equilibrium, it might happen that the chances of molecular agitation would collect all the oxygen molecules to one side and all the nitrogen to the other; all that then need be done to keep the two gases separate would be to close a tap in the connecting tube. The kinetic theory enables us to calculate the time T (which will be very long if the number of molecules is large) during which spontaneous separation of this kind will have an even chance of occurring. Consider now a very large number of similar pairs of globes. During each lapse of time T, whatever time has elapsed [1] already, spontaneous separation will occur in half the pairs of globes still effective; the variation law is the same as for radioactive elements.

The above illustration makes it clear, in my opinion, that in each atomic nucleus (comparable with the gaseous mixture that fills one of our pairs of globes) a statistical equilibrium must be set up between a large number of irregularly varying parameters, as in the case of a gas in equilibrium, or of light filling an isothermal enclosure.

When, *by chance*, certain conditions that are as yet unknown are satisfied within the complex nucleus, a fundamental upheaval occurs, the atom exploding like a charge of dynamite that can be detonated by a small spark.[2]

I need scarcely point out that the law of chance found for the radium and thorium emanations is the general law of

[1] It makes no difference whether or not separation had, at a given instant, say after one hour, been nearly complete for any pair; in general, such a state of affairs cannot persist for very long, since a return to a state of mixture is much more probable for a partially separated system (see Borel, Le Hasard, p. 174).

[2] The necessary conditions for decomposition may, however, depend upon phenomena external to the atom, in the form of a penetrating radiation (ultra-X) coming from the earth and providing the energy necessary for the disruption of a system normally very stable. I have recently developed this hypothesis (Ann. de Phys., 1919). (See Appendix.)

GENESIS AND DESTRUCTION OF ATOMS

atomic disintegration. To each radioactive element corresponds a definite *period* or time during which half of any measurable mass of the element undergoes transmutation. This period is about 1,800 years for radium (Boltwood), so that if a tube containing 2 grammes of it were sealed up in 1900 there would not be more than 1 gramme of radium in the tube in the year 3700, together with ·3 gramme of helium and 1·7 grammes of lead. As may be shown by a simple calculation, this may be also expressed by the statement that very nearly $1·2 \times 10^{-11}$ part of any given mass of radium disappears per second.

111.—RADIOACTIVE SERIES.—It has been possible (as was

(5×10^9 years) . .	Uranium. ↓
(24 days) . . .	Uranium X_1 + Helium. ↓
(1·5 minutes) . .	Uranium$_2$ X + β. ↓
(2×10^6 years) . .	Uranium II + β. ↓
(100,000 years) . .	Ionium + Helium. ↓
(1,800 years) . .	Radium + Helium. ↓
(3·85 days) . .	Niton + Helium. ↓
(3 minutes) . .	Radium A + Helium. ↓
(27 minutes) . .	Radium B + Helium. ↓
(19 minutes) . .	Radium C + Helium. ↓
(very short) . .	Radium C' + β. ↓
(16 years) . . .	Radium D + Helium. ↓
(5 days) . . .	Radium E + β. ↓
(136 days) . .	Polonium + β. ↓
(∞) . . .	Lead + Helium.

done for niton before it was isolated) to characterise by their periods no less than thirty new simple substances, derived from uranium and thorium by successive transmutations.[1] One of these periods is no more than the one twenty-fifth of a second (and there are certainly others even shorter); others exceed 1,000,000,000 years. In the table above are shown the periods T for a series of elements derived from uranium by successive internal decompositions or rearrangements.

Bifurcations are possible, *side chains* being formed.[1] In other words, the same atom may undergo, according to which of two critical internal configurations *happens* to occur first, one or another kind of transmutation. We may suppose that if, during the same time, a uranium atom had 9 chances out of 10 of undergoing the rearrangement that gives uranium X, and 1 chance in 10 of undergoing another that would give actinium, the whole of any measurable mass of uranium would be transformed $\frac{9}{10}$ into uranium X and $\frac{1}{10}$ into actinium.

It will be noticed that helium (which undoubtedly has a very stable nucleus) is a *frequent* product of atomic disintegration.[2] This perhaps explains why many of the differences between atomic weights (lithium and boron, carbon and oxygen, fluorine and sodium, etc.) are exactly equal to 4, the atomic weight of helium.

It is to be presumed that other chains of transmutations may show smaller differences equal to the atomic weights of hydrogen or nebulium. Moreover, a radioactive element, though classed as emitting only β and γ rays, might very well project atoms heavier than the helium atom (copper, for instance), without our becoming aware of it, for reasons that will be apparent later.

112.—Cosmogony.—If the inverse phenomenon is possible and radioactive atoms can be regenerated, the process must take place at the centres of stars, where the temperature and pressure is enormous and favours reciprocal

---

[1] One such chain, which leads to actinium, perhaps starts from uranium II.

[2] In terms of Prout's hypothesis, we may regard the helium nucleus as made up of four hydrogen nuclei kept together by two electrons, the α particle thus having a charge 2 (Rutherford). The stability of such an assemblage is discussed in the Appendix.

GENESIS AND DESTRUCTION OF ATOMS

penetration between atomic nuclei, accompanied by energy absorption.[1]

The high value given by analysis for the mean radioactivity of the earth's crust appears to me to afford a strong presumption in favour of this hypothesis. If radioactive atoms were equally abundant near the centre, the earth would be more than 100 times more radioactive than is sufficient to account for the preservation of its central heat. It has therefore been suggested that such atoms are present only in the superficial layers. This view appears to me to be incorrect, for the radioactive atoms, being very heavy, ought on the contrary to accumulate *enormously* at the centre. We are therefore forced to accept a *very rapid* rate of cooling for the earth, unless we assume that in the deeper layers a highly endothermic formation of such atoms occurs.[2]

113.—ATOMIC PROJECTILES.—The penetration of α rays into matter gives us important information about the atoms and the singular properties they may acquire when protected at the enormous speeds possessed by these rays.

The essential fact is that α rays pass in straight and sharply defined lines, without noticeable diffusion, through layers of air several centimetres thick, and through homogeneous thin sheets of aluminium and of mica up to four or five hundredths of a millimetre in thickness.

Now, taking the atomic diameter in the sense employed in the kinetic theory (diameter of impact), we find that the atoms in aluminium or mica are as closely packed together as the shot in a pile of shot. It cannot be supposed that the helium projectiles pass through the interstices, and we must assume that they pierce the atoms, or more accurately the casings (para. 95), that protect the atoms from molecular impacts. It is easily shown, from the density of aluminium, that each α projectile pierces about 100,000 aluminium

[1] This hypothesis, which Mme. Curie put forward at the same time as myself (1912), certainly expresses the attitude of many physicists.

[2] Since the above was written the possibility of a more precise theory has occurred to me. It is suggested that heavy atoms are produced, with liberation of energy, by the condensation of lighter atoms. This condensation occurs in the stars and is the origin of stellar radiation (Annales de Physique, 1919, and Revue de Mois, 1920). (See Appendix.)

atoms before it is stopped. This will not seem so surprising if it is remembered that the initial energy of such a projectile is more than 100,000,000 times greater than that of a molecule in ordinary thermal agitation. Finally, the thin metallic sheets exposed to this bombardment do not appear to be altered.

Extrapolation to the case of any kind of atom whatever is certainly permissible, and we may picture two atoms colliding at sufficiently high speeds as passing through each other without mutual effect.[1] This becomes comprehensible when we remember what has been said as to the extreme smallness of the volume actually occupied by the material part of the atom (para. 94). If a star happened to be impelled towards the solar system, regarded as bounded by the orbit of Neptune, the chances are small that it would hit the sun itself. If, moreover, the relative motions of the star and the sun were sufficiently rapid, the forces of attraction would not have time to do any reasonable amount of work and neither star nor sun would be deviated perceptibly from their courses. Similarly, the extreme smallness of the atomic nucleus certainly makes actual impact between nuclei extremely rare. But a few peripheral corpuscles that offer less resistance to being set in motion may get detached, with the result that the projectile leaves a train of ions behind it.

In consequence of the ionisation thus produced, α rays gradually lose their velocity as they pass through matter. The surprising fact has been established that all their characteristic properties cease to be shown when their velocity falls to a certain critical value, which, however, is still very high (more than 6,000 kilometres per second).

Consider a minute speck of polonium in air ; the α rays emitted suddenly cease to have effect on reaching the circumference of a sphere of radius 3·86 centimetres with the grain as centre. About a speck of radium *in radioactive equilibrium* (containing, that is to say, the limiting proportions of the successive products of its disintegration),

[1] A rifle bullet *moving sufficiently rapidly* would pass through a man without hurting him.

GENESIS AND DESTRUCTION OF ATOMS

it is possible to trace five sharply defined concentric spheres with radii lying between 3 and 7 centimetres.[1]

It was at first supposed that this fact established a difference in nature between α rays and the positive rays from a Crookes' tube, wherein the velocity is only a few hundred kilometres per second, although the particles travel in straight lines for several decimetres. But a distance of several decimetres in a Crookes' tube is not equivalent to a hundredth of a millimetre in ordinary air. It is now held, therefore, quite simply, that the penetrating power, being a function of velocity, falls off very rapidly when the velocity falls below a so-called critical value (ill-defined), so that an atomic projectile that cannot do more than, say, 5,000 kilometres per second cannot pass through more than a quarter of a millimetre of air. Moreover, towards the end of its path ionisation becomes intense and diffusion considerable, until finally the projectile gets very considerably slowed down, no longer breaks through the atomic casings, and rebounds from them like an ordinary molecule.

It is now apparent why I took occasion to point out (para. 111) that if an atomic explosion were to project a sufficiently heavy atom of some common element we should not be able to perceive it. In such cases a *masked transmutation* would occur. For explosive energies of the order of magnitude established up to the present only the lighter atoms could acquire sufficient velocity and energy to give them a noticeable path in air ; a copper atom, for instance, could not be detected.

COUNTING ATOMS.

114.—SCINTILLATIONS : THE CHARGE ON THE α PROJECTILE.—Sir William Crookes discovered that the phosphorescence excited by the α rays in substances that stop them is resolved under the magnifying glass into separate

---

[1] In minerals the circumferences are microscopic and appear as small round spots (pleochroic halos), which are observed about minute feebly radioactive crystals imbedded in certain micas. The extent of the blackening produced by activity of a known kind enabled Joly and Rutherford to estimate the time taken by the halo to form as several hundred million years.

scintillations, fugitive starlike points of light that are extinguished as soon as they are kindled. They may be seen continually appearing and disappearing all over the screen that receives the stream of projectiles. Crookes at once suggested that each scintillation marks the point of arrival of one projectile and thus enables us to perceive, for the first time, the individual effect of a single atom. Similarly, although we may not see a shell, we can perceive the conflagration that it kindles when it is stopped.

Rutherford, moreover, had measured, in a Faraday cylinder, the positive charge q radiated per second in the form of α rays from a given mass of polonium, and (by measuring the conductivity of the gas) had determined the positive and negative charges $+ Q - Q$ that the same rays liberate in ionising the atoms they pass through before being stopped in air. In this way he had found that the liberated charges Q were equal to very nearly 100,000 times (94,000 times) the charge q carried by the projectiles.

Combining the two processes, Regener determined the molecular magnitudes in a new way. He counted one by one the scintillations produced within a given angle by a given polonium preparation and from the result calculated the total number of α particles emitted per second by that preparation (1,800 in point of fact). He found, moreover, that in one second these particles liberate ·136 electrostatic unit of each sign in air. This gives the charge $\dfrac{\cdot 136}{1,800 \times 94,000}$, or 8×10^{-10} for each α particle. Since the α projectile carries twice the elementary charge, the value of the latter must be 4×10^{-10}, which agrees well with the other determinations.

115.—Electrical Methods of Counting.—In spite of this agreement, it might still be doubted whether the scintillations are exactly equal in number to the number of projectiles. Rutherford and Geiger extended and consolidated Regener's brilliant work and devised a second extraordinarily ingenious method for counting the projectiles.

In their apparatus the α projectiles start from a *thin*

GENESIS AND DESTRUCTION OF ATOMS

radioactive layer of known surface (radium C) and are filtered through a mica diaphragm (thin enough for all of them to pass through). They then enter a gas at low pressure between two plates at different potentials, one being connected with a sensitive electrometer. In the gas each projectile produces a train of ions which move, according to their sign, towards one or other of the electrodes.

If the pressure is sufficiently low and the potential difference sufficiently high, it becomes possible for each ion to acquire a velocity in the interval between two molecular impacts fast enough to split up the molecules it meets into ions, which become ionising centres in their turn.[1] This multiplies quite a thousandfold the discharge that would be caused by those ions only that were produced by the projectiles directly. The discharge is thus made large enough to be detected by a noticeable deflection of the electrometer needle.[2] Under these conditions, the radioactive source being sufficiently far removed and the α radiation that it sends between the two plates being limited in amount by its passage through a small aperture, the movements of the electrometer needle are seen to take place in *distinct jerks* irregularly distributed in time (from two to five per minute). This very clearly demonstrates the granular structure of the radiation.

The jerks can be counted with rather greater precision than the scintillations, and the numbers obtained by the two methods are equal. Rutherford found that 1 gramme of radium in equilibrium (with its disintegration products) emits 136,000,000,000 helium atoms per second, which means that radium by itself produces 34,000,000,000 (3.4×10^{10}) projectiles per second.

Omitting Regener's intermediate step, Rutherford and Geiger then allowed α projectiles, emanating from a thin radioactive layer and n in number, determined as above, to fall within a Faraday cylinder (the negative β particles, being much more readily deviated by a magnet, were

---
[1] This phenomenon was discovered by Townsend and is the basis of the present explanation of the mechanism of the disruptive discharge (electric spark).
[2] Rapid return of the needle to zero is assured by making the insulation imperfect.

removed by a powerful magnetic field). The quotient $\frac{q}{n}$ of the positive charge q that gets into the cylinder by the number of projectiles n gives the charge $9 \cdot 3 \times 10^{-10}$ borne by a projectile, which gives $4 \cdot 65 \times 10^{-10}$ for the elementary charge e, and

$$62 \times 10^{22}$$

for Avogadro's number, with an error of probably not more than 10 per cent.[1]

116.—THE NUMBER OF ATOMS THAT GO TO MAKE UP A KNOWN VOLUME OF HELIUM.—Since we can count the α projectiles emitted in a second by a radioactive body, we know how many atoms there are in the mass of helium produced during that time. If we determine that mass or the volume it occupies at a given temperature and pressure, we shall obtain the mass of the helium atom directly. The difficulty, by no means small, is to collect all the helium and to prevent its contamination with other gases.

Measurements carried out by Sir James Dewar and subsequently improved upon by Boltwood and Rutherford, indicate that 156 cubic millimetres are liberated annually per gramme of radium. Allowing for the disintegration products present with the radium, this gives 39 cubic millimetres for the pure radium alone. Since it projects 34,000,000,000 helium atoms per second, we get $34 \times 86,400 \times 365$ thousand million molecules in that volume. The number of monatomic molecules N of helium that occupy

---

[1] Regener has recently carried out determinations of this kind with α rays from polonium by counting the scintillations produced on a homogeneous flake of diamond. His determination of the charge q seems to me, however, to involve an uncertain factor, and a short discussion will be of interest.

In this method it is implicitly assumed that *the whole of the charge* registered by the receiver is carried by α projectiles. Now, the explosion that propels an α projectile in one direction also propels the rest of the radioactive atom, a, in the opposite direction. These a rays, which have scarcely any penetrating power cannot have any effect in Rutherford's apparatus (in which a thin plate separates the active body from the receiver). But in Regener's experiment they may exert their influence (the ends of the apparatus being open and no thin screen being used). For it is possible that these a rays do not produce scintillations; it is probable that they are positively charged (like all violently projected atoms) and that they carry two positive charges, like helium. In short, the value $4 \cdot 8 \times 10^{-10}$ obtained cannot be regarded as *certain*.

22,400 cubic centimetres, and which therefore make up a gramme molecule, is thus

$$\frac{34 \times 86{,}400 \times 365 \times 22{,}400}{\cdot 039} \times 10^9 \text{ or } 62 \times 10^{22}.$$

Mme. Curie and Debierne subsequently carried out a similar determination of the quantity of helium liberated by polonium.[1]

Projectiles were counted, as in Rutherford and Geiger's experiments, by the scintillation method, and by the method of "electrometer jerks." The latter, made to occur at considerable intervals (one per minute) so that they should not overlap, were recorded on a ribbon, each jerk being indicated by a small denticulation in a continuous line.

Fig. 14.

The denticulations could then be counted at leisure [2] (Fig. 14). The volume of helium liberated was ·58 cubic millimetre. This series of experiments gives for N the value

$$65 \times 10^{22}$$

which is in remarkable agreement with the values already obtained.

117.—THE NUMBER OF ATOMS THAT MAKE UP A KNOWN MASS OF RADIUM.—The number of projectiles emitted gives the number of generative atoms that disappear as well as the number of helium atoms that make their appearance. If, therefore, we have any means of finding out what fraction

[1] The choice of polonium offers many advantages, because the radioactive phenomena in connection with it are less complex, polonium being the end product of its radioactive series (only one transmutation occurs, into helium), and because, no gaseous emanation being produced in the space where the radioactive material is mounted, the number of α projectiles that penetrate into the glass is negligible; in this way the difficulties involved in the removal of helium occluded in the glass are avoided.

[2] Taken, for convenience in printing, from some later work of Geiger and Rutherford, in which the projectiles from radium were counted by this method with very great accuracy.

of a gramme atom of the generative body has disappeared, we can obtain at once the mass of the atom of that body and hence the other molecular magnitudes.

All the necessary data are available in the case of radium; its gramme atom is known to be 226·5 grammes, and the loss in α projectiles is $3\cdot4 \times 10^{10}$ per gramme. The gramme atom therefore emits $226\cdot5 \times 3\cdot4 \times 10^{10}$ α projectiles per second. We know, moreover (para. 110), that out of N radium atoms $N \times 1\cdot09 \times 10^{-11}$ disappear per second, which gives N from the equation

$$226\cdot5 \times 3\cdot4 \times 10^{10} = N \times 1\cdot09 \times 10^{-11};$$

we thus get

$$N = 71 \times 10^{22}.$$

118.—The Kinetic Energy of an α Projectile.—If we know, as is the case with radium, the kinetic energy and speed of the α projectiles, we can obtain, in yet another way, the mass of the helium atom and the molecular magnitudes.

The kinetic energy to within a few per cent. (due to the penetrating β and γ rays) is equivalent to the heat continually liberated (Curie). Let u_1, u_2, u_3, u_4 be the initial velocities (determined by Rutherford) of the four series of α projectiles emitted by radium in radioactive equilibrium. Since radium liberates 130 calories per gramme per hour (3,600 seconds), and since the mass of one helium atom is $\frac{4}{N}$, we have, very nearly,

$$\frac{1}{2} \times \frac{4}{N} \times 3\cdot4 \times 10^{10}[u_1^2 + u_2^2 + u_3^2 + u_4^2] = \frac{130 \times 4\cdot18 \times 10^7}{3,600},$$

or a value for N of nearly 60×10^{22}.

The individual energy of an α particle is of the order of a hundred-thousandth of an erg.

119.—The Path of each Atomic Projectile can be made Visible.—Thanks to the scintillations produced, we are able to perceive the stoppage of each of the helium atoms that constitute the α rays. But the path followed by each

GENESIS AND DESTRUCTION OF ATOMS

atom is nevertheless invisible, and we only know that it is approximately rectilinear (since the α rays scarcely diffuse at all), and that it must be marked by a train of ions, liberated from the atoms passed through. Now, in an atmosphere saturated with water vapour, each ion can act as the nucleus of a visible drop (para. 99), and C. T. R. Wilson, who discovered this phenomenon, has made use of it, in a most ingenious manner, to demonstrate the path as a visible streak.[1]

A minute radioactive speck, placed at the end of a fine wire, is introduced into an enclosed space saturated with water vapour. A sudden expansion increases the volume and produces supersaturation by cooling. At very nearly the same instant a spark is produced and lights up the enclosure. In the form of white rectilinear streaks starting from the active granule rows of droplets can be seen (and photographed) along the paths followed by the few particles emitted after the expansion and before the illumination of the vessel (Fig. 15).

Fig. 15.

Closer examination, however, shows that the trajectories are not rigorously straight, but bend noticeably during the last few millimetres of their path, and even show sharp angles (several are visible in the figure). Each time the atomic projectile passes through an atom it undergoes a deviation, very slight, but nevertheless not absolutely negligible; these deviations, which act cumulatively and in opposition to one another quite irregularly, explain the observed tendency to curve. Finally, in very exceptional cases (owing to the extreme smallness of the atomic nuclei)

---

[1] Proc. Roy. Soc. A., Vol. LXXXVII., 1913.

214   ATOMS

it happens that the nucleus into which almost all the mass of the projectile is condensed strikes the nucleus of another atom ; a considerable deviation is then suddenly produced. At the same time, the nucleus that has been struck receives an impulse sufficiently intense to make it become, in its turn, an ionising projectile, with a trajectory that, although very short, is nevertheless recorded quite clearly on the plate as a kind of spur.[1]

Finally, C. T. R. Wilson has succeeded in making visible, by the condensation of water droplets, the path followed by an ionising *corpuscle* (β rays and kathode rays). The phenomenon is particularly interesting in the case of

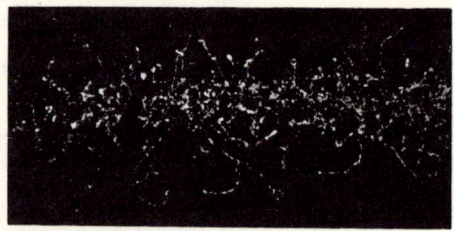

FIG. 16.

the secondary rays (having small penetrating power and diffuse trajectories) produced by the emission of corpuscles from atoms struck by γ rays or X rays. Curving of the trajectory is then very marked. Moreover, since their ionising power is less than for α rays, the droplets appear separated from each other and give a visible indication of each ionising impact. Fig. 16, which is a photograph of the trajectory in air of a pencil of X rays, shows that the primary ionisation is of very little importance and that nearly all the ions are produced along the curvilinear

---

[1] In this we have, I think, a means of estimating the relative dimensions of the atomic nucleus and the atom. We have only to find how many single impacts occur on the average in the trajectory of an α particle on passing through an approximately known number of atoms (p. 205). An examination of Wilson's photographs seems to me to indicate (very roughly) that one nuclear impact occurs for every million atoms traversed ; *the diameter of the atomic nucleus should therefore be about a thousand times less than that of the atom.*

GENESIS AND DESTRUCTION OF ATOMS

trajectories of the various secondary rays produced by the primary ionisation.

The beauty of these brilliant experiments needs no comment.

CONCLUSIONS.

120.—THE AGREEMENT BETWEEN THE VARIOUS DETERMINATIONS.—In concluding this study, a review of various phenomena that have yielded values for the molecular magnitude enables us to draw up the following table:—

Phenomena observed.[1]	$\dfrac{N}{10^{22}}$
Viscosity of gases (kinetic theory)	62 (?)
Vertical distribution in dilute emulsions	68
Vertical distribution in concentrated emulsions	60
Brownian movement { *Displacements*	64
Brownian movement { Rotations	65
Brownian movement { Diffusion	69
Density fluctuation in concentrated emulsions	60
Critical opalescence	75
Blueness of the sky	65
Diffusion of light in argon	69
Black body spectrum	61
Charge as microscopic particles	61 (?)
Radioactivity { Projected charges	62
Radioactivity { *Helium produced*	66
Radioactivity { Radium lost	64
Radioactivity { Energy radiated	60

Our wonder is aroused at the very remarkable agreement found between values derived from the consideration of such widely different phenomena. Seeing that not only is the same magnitude obtained by each method when the

[1] Methods by which it may be hoped, in the future, to obtain results of great precision are given in italics.

conditions under which it is applied are varied as much as possible, but that the numbers thus established also agree among themselves, without discrepancy, for all the methods employed, the real existence of the molecule is given a probability bordering on certainty.

Yet, however strongly we may feel impelled to accept the existence of molecules and atoms, we ought always to be able to express visible reality without appealing to elements that are still invisible. And indeed it is not very difficult to do so. We have but to eliminate the constant N between the 13 equations that have been used to determine it to obtain 12 equations in which only realities directly perceptible occur. These equations express fundamental connections between the phenomena, at first sight completely independent, of gaseous viscosity, the Brownian movement, the blueness of the sky, black body spectra, and radioactivity.

For instance, by eliminating the molecular constant between the equations for black radiation and diffusion by Brownian movement, an expression is obtained that enables us to predict the rate of diffusion of spherules 1 micron in diameter in water at ordinary temperatures, if the intensity of the yellow light in the radiation issuing from the mouth of a furnace containing molten iron has been measured. Consequently the physicist who carries out observations on furnace temperatures will be in a position to check an error in the observation of the microscopic dots in emulsions! And this without the necessity of referring to molecules.

But we must not, under the pretence of gain of accuracy, make the mistake of employing molecular constants in formulating laws that could not have been obtained without their aid. In so doing we should not be removing the support from a thriving plant that no longer needed it; we should be cutting the roots that nourish it and make it grow.

The atomic theory has triumphed. Its opponents, which until recently were numerous, have been convinced and have abandoned one after the other the sceptical position that was for a long time legitimate and no doubt useful. Equilibrium between the instincts towards caution and towards boldness is necessary to the slow progress of human science ; the conflict between them will henceforth be waged in other realms of thought.

But in achieving this victory we see that all the definiteness and finality of the original theory has vanished. Atoms are no longer eternal indivisible entities, setting a limit to the possible by their irreducible simplicity ; inconceivably minute though they be, we are beginning to see in them a vast host of new worlds. In the same way the astronomer is discovering, beyond the familiar skies, dark abysses that the light from dim star clouds lost in space takes æons to span. The feeble light from Milky Ways immeasurably distant tells of the fiery life of a million giant stars. Nature reveals the same wide grandeur in the atom and the nebula, and each new aid to knowledge shows her vaster and more diverse, more fruitful and more unexpected, and, above all, unfathomably immense.

APPENDIX, 1921

Corpuscular Swarms.—We have seen that the atom is probably composed of a nucleus of enormous density, a positive sun about which gravitates a swarm of negative planets ; these planets are identical among themselves, and with the negative electrons or corpuscles which, when moving outside of atoms, are the kathode or β rays. Moreover, the circulation of this swarm cannot radiate energy, otherwise the atom would undergo a progressive change in properties, while, at the same time, it would contract until the corpuscles had reunited with the nucleus.

Since the atom is neutral, the nuclear charge will be as many times that of a corpuscle as there are planets in the swarm. This nuclear charge, which determines the number of planets, at the same time determines the configuration and all the characteristics of the swarm. So that if two nuclei, otherwise different, bear the same charge, the corresponding atoms will have identical corpuscular swarms and will behave in the same way in regard to those properties which are determined by the swarm (spectrum lines, cohesion, chemical affinity). In spite of their different masses, such atoms will thus occupy *the same place* in any chemical classification, and for that reason they are called isotopes (Soddy). On the other hand, two atoms having different nuclear charges will differ chemically, even though they have exactly the same weight (for example, *bismuth* and the *lead* derived from thorium each have the atomic weight 208).

It has been found possible to determine this nuclear charge, or, which comes to the same thing, the number of corpuscles in the swarm. Two very different methods have been employed. The first depends upon measurements of the diffusion of X rays, which is probably the sum of the diffusions due separately to the electrons which affect the waves during their passage (Barkla). The second depends on measurements of the lateral dispersion undergone by X rays on passing through a piece of thin metal foil ; the rectilinear movement of some of the rays is suddenly bent (as in C. T. R. Wilson's experiments). The dispersion thus measured is quantitatively explained if the centres of the atoms encountered by the X rays are identical positive charges arranged in a regular order, provided that these nuclear charges have a fixed value; these values can be calculated from the dispersion with a precision which reaches 1 part in 100 (Rutherford).

By each of the above methods it is found that the number of corpuscles in the swarm is roughly equal to half the atomic weight, and, more exactly, equal to the sequence number (atomic number, not always quite definite) which Mendélejeff's series indicated for the atom. It thus becomes probable that the nuclear charge increases unit by unit, without gaps, one corpuscle entering the swarm at each step. The *sequence number* (atomic number) of each element in the series thus obtained will be equal to the number of corpuscles that gravitate about its nucleus (Van der Brock). *Moseley's* [1] *Law* fixes without ambiguity for each element this sequence number, the ultimate significance of which we now understand.

Every element when bombarded with kathode rays emits, as we know, X rays. These X rays, when analysed spectroscopically by " reflection " from the recticular planes of a crystal (Bragg's method), give, instead of a continuous spectrum, sensibly monochromatic rays, which are the " characteristic X rays " of the element. The more penetrating of these radiations, the so-called K radiation, which is the most penetrating of the various kinds of light that we can cause an atom to emit,[2] can be resolved into a group of narrow " rays "; this group is clearly recognisable for the various elements, the rays being the more penetrating the higher the element is in the series.

Let us take a particular one of these rays that make up the K radiation and plot a point for each element having the sequence number as abscissa and for ordinate the square root of the frequency of the ray. Moseley's Law expresses the fact that such points lie *rigorously* on a straight line,[3] provided that in the sequence of atomic weights the alterations already suggested by Mendélejeff's Law are made (tellurium before iodine, for example), and provided that five specified places are left vacant (43, 61, 75, 85, 87) between hydrogen (atomic number 1) and uranium (atomic number 92). Only five elements (several isotopes of each being, of

[1] Killed at the Dardanelles.
[2] About 50,000 times more penetrating than visible light.
[3] Which passes a little below the origin; in other words, the frequency is approximately proportional to the square of the sequence (atomic) number.

course, possible) thus remain to be discovered in this interval.

Let us write the series of the elements in this order, which is now fixed with certainty. We know that as we proceed along the series we shall find elements analogous to elements already classified and in the same order. It is probable that the peripheral configuration of the corpuscular swarm is repeated in very nearly the same manner, so that the addition of new corpuscles, if they remain superficial, produces in the two cases analogous peripheral changes. But if, under the influence of the deeper layers, which are more and more complex, the added electron is forced into the swarm instead of remaining on the surface, the periphery, as well as the properties of the atom, will be little changed (rare earths ?). The table given at the end of the Appendix shows the classification of, and the analogies between, the elements arranged according to their order, and summarises the precise form in which we may nowadays accept Mendélejeff's Law.

Bohr's Theory.—We have seen that, since an atom possesses and retains definite properties, its corpuscular swarm should have a definite and fixed configuration. But, always on the assumption that Coulomb's Law is obeyed, classical mechanics would allow an infinite series of configurations up to the most stable limit corresponding to the reunion of the planetary electrons with the central nucleus.

For example, considering only the hydrogen atom, we could imagine an infinite number of elliptical orbits for the corpuscle gradually closing in upon the nucleus.

We must therefore conclude that laws of an entirely new kind must intervene in the atomic world, to choose between all these configurations. Bohr's remarkable theory, which has received several remarkable confirmations, has already thrown some light on this problem.

Bohr assumes, in the first place, that there is a discontinuous series of non-radiating stationary states possible for the corpuscular swarm, selected from among these allowed, by Coulomb's Law and the classical mechanics.

This series has been calculated only for an atom with one corpuscle (hydrogen, or ionised helium). The possible orbits must, according to Bohr, be such that their area, multiplied by the mass of the planet,[1] gives a whole number multiple

---

[1] In other words, the moment of the quantity of motion with respect to the centre of attraction.

of $\frac{h}{2\pi}$ (h being Planck's constant). For circular orbits we thus find that the diameters are proportional to the squares of successive whole numbers. Of these systems the most stable, and, strictly speaking, the only stable, configuration is the one with the smallest orbit, in which case the internal energy of the atom is a minimum. In the case of hydrogen the reader will readily find the value 10^{-4} μ for the diameter of this minimum orbit. This is more or less the value predicted above (para. 95).

Bohr supposes further that if the swarm happens not to be in its most stable state, it soon passes " spontaneously " into a more stable one ; it must then radiate energy equal to $\Delta\omega$, the difference between the energies of the two states. Bohr assumes that this radiation takes the form of monochromatic light of frequency $\frac{\Delta\omega}{h}$. In this manner he has calculated quantitatively and with precision the series of frequencies of the rays from hydrogen, not only for the well-known Balmer series but also for the infra-red (Paschen), and for an ultra-violet series discovered by Hyman, as a matter of fact after the appearance of Bohr's work. Even more striking confirmation of the theory has been obtained for helium, which, when it loses a corpuscle by ionisation, becomes a kind of hydrogen containing a nucleus of charge 2 with a single corpuscle gravitating about it.

The above sketch, though brief, will at least enable us to understand how the corpuscular swarm, in its most stable state, gravitates at an appreciable and definite distance from the nucleus ; a closer permanent approach between the corpuscles and the nucleus is impossible. We shall now turn to certain systems that are in some respects the inverse of the above, and in which large masses are gravitating ; the dimensions of these systems pass suddenly from one order of magnitude to another very much lower.

Nuclei.—Knowing the mass of a corpuscle and the number of corpuscles in the swarm, we now know accurately, as was suspected above (para. 94), that the planetary swarm, which is so important from the point of view of the properties of the element (on account of the attractions and repulsions that it exerts), has a total mass negligible compared with that of the nucleus in which to within a ten-thousandth part the material part of the atom is found to be condensed.

Let us now turn our attention to this dense, unimaginably small nucleus. We are given the clue to its structure by a fact of capital importance, namely, the verification of Prout's Law. Each atom, or, rather, each nucleus, contains the hydrogen atom or nucleus a whole number of times. If we represent the hydrogen nucleus or positive protoatom by h and the corpuscle or negative protoatom by β we may say that all matter is made up of protoatoms h and β.

Names are required for these two fundamental constituents. The negative protoatom may be called the *corpuscle*[1] and the positive protoatom the *nucleon*.[2]

Thus the helium nucleus α, to which we may give the formula $h_4\beta_2$, is made up of four h nucleons, cemented together by two β corpuscles, so that the nuclear charge is two units. It is readily seen that two corpuscles, placed at two opposite apices of a tetrahedron, can keep in stable dynamic equilibrium about the axis joining the corpuscles, four h nucleons placed at the four remaining apices, the forces of inertia balancing the electrostatic forces. *Here heavy masses move about light masses*, and, supposing the condition required by Bohr to hold, this should impose on the system a diameter about a thousand times smaller than that of the hydrogen atom. This is not small enough, the dispersion of α particles (Rutherford) requiring a diameter about twenty times smaller still. As a preliminary explanation it is, however, sufficiently remarkable for us to conclude that the laws of quanta hold in the nucleus.[3]

The loss in potential energy $\Delta\omega$ that occurs when 4 nucleons and 2 corpuscles condense to form a helium nucleus should reappear in the form of radiation, and, furthermore, according to the Einstein relativity principle, a loss of energy equal to $\dfrac{\Delta\omega}{C^2}$ (C^2 being the square of the velocity of light) should take place. In actual figures, the weight of 4 gramatoms of hydrogen exceeds that of a gramatom of helium by 3 centigrammes. This corresponds, according to the Einstein formula, to the liberation of the enormous amount of energy, 7×10^{12} calories. I have drawn attention to this point and pointed out that, assuming that the primary nebula was composed of hydrogen, such an evolution of energy would

---

[1] Suggested by Millikan, and, in his opinion, to be preferred to *negative electron*, which has also been proposed.

[2] Suggested by P. Auger. *Proton* has also been suggested.

[3] This theory was put forward by Sommerfeld, who has much extended and perfected Bohr's theory.

APPENDIX, 1921

supply solar radiation at its present actual rate of evolution for at least 10^{11} years.[1]

Because of this enormous liberation of energy, which must be restored to the atom before it can be disintegrated again, the helium nucleus will be very stable ; it is found as a subnucleus frequently present in heavier nuclei, as is proved by the many radioactive transformations that take place with emission of α particles. An additional proof is given by the fact that in the list of atomic weights we often find a difference in weight of 4 units between atoms for which the difference in atomic number (or nuclear charge) is 2 units.

We may ask whether, when the weight of an atom is a multiple of 4 (such as argon, of weight 40), its nucleus is not made up entirely of helium nuclei. As a matter of fact, in the case of argon, of atomic weight 40 and atomic number 18, this is impossible, because the charges on 10 helium subnuclei would require an atomic number 20. The presumption is, however, strengthened when at the same time the atomic weight is $4n$ and the atomic number $2n$. This can only happen for elements of atomic number less than 20 ; it does, however, occur with the series of elements in the middle of that range (such as carbon and sulphur) and cannot be put down as mere chance.

We must therefore conclude that α nuclei can group themselves together, in spite of the repulsions due to the charges on them. We have, as a matter of fact, already assumed an electrical " cohesion " at very small distances in order to reconcile ourselves to the fact that the electron, which cannot be a mathematical point, does not fly to pieces under the influence of the reciprocal repulsions of its parts.

Another stable nucleus, present as a sub-nucleus in the heavier nuclei, is possibly the *nebulium*, of atomic weight 3, discovered by Buisson and Fabry in the nebula of Orion. And I am inclined to the opinion that this nebulium is composed of 3 nucleons gravitating about 2 corpuscles and is an isotope of hydrogen (and not of helium). A presumption in favour of this view seems to me to come from the fact that the difference in atomic weight is often 3 between atoms which differ by 1 in atomic number (or nuclear charge). More precisely, and dealing with known isotopes, I find 30

---

[1] This would provide the solution to the problem of the origin of the heat of the sun (J. Perrin, "Maliève et Lumière." *Ann. de Phys.*, 1919, and *Revue du Mois*, 1920).

elements for which the atomic weight is $(4n + 3m)$, the atomic number being $(2n + m)$, (n and m being whole numbers and not zero).

The constitution of certain light nuclei, in which the nucleon perhaps functions as an independent sub-nucleus, remains to be made out. On the other hand, the emission of β rays from radioactive bodies would lead one to suppose that, at any rate in the centres of certain heavy nuclei, the corpuscles themselves act as sub-nuclei.

The nucleons, which exist independently within the nucleus or hidden in sub-nuclei, can be expelled by means of a suitable shock. Rutherford has succeeded in bringing about the emission of hydrogen atoms from atoms of B, N, F, Na, Al, P, by bombarding them with α particles from radium C. The liberation of nucleons is shown by the scintillations produced on a zinc sulphide screen at 'distances considerably greater than those to which nuclei could travel in chaotic motion according to the laws of impact. Nucleons are, in fact, projected backwards as well as forwards with an amount of energy that, for Al or P, reaches 1·3 times that of the projectile that causes their expulsion (I presume that this projectile combines with the nucleus that it hits, one nucleon being expelled and energy radiated away).

It will be noticed that the above preliminary sketch of the structure and internal properties of the nucleus gives no suggestion of the possibility of those spontaneous explosions that at the present time are assumed in order to explain radioactivity (para. 110). And, as a matter of fact, I believe that, far from being a kind of charge of dynamite that is exploded by chance, an atom of, say, radium is a very stable system which can be disintegrated only when a considerable amount of energy is given to it from outside. This energy could be provided by some kind of radiation, which influences the nucleus *according to the ordinary laws of quanta;* from this we are able to calculate its frequency. I find in this way that the light thus active stands very nearly in the same relation to X rays as the latter do to ordinary light. These *ultra X rays* cannot possibly come from the sun or from any direction fixed with respect to the stars because radioactivity is the same by day as at night (Curie). They may, however, be thrown out by the hot central region of the earth itself, passing through the terrestrial shell without much loss in energy. The radioactivity of elements would in that case not be invariable. And the primary phenomenon that gives rise to ultra X rays, and to all stellar radiation,

APPENDIX, 1921

would be the progressive condensation, occurring at the centres of the stars, of protoatoms into nuclei of progressively increasing weight.

LIST OF ELEMENTS

(Atomic Numbers, Names, Symbols, Atomic Weights, Isotopes,[1] etc.).

1. *Hydrogen*, H (1·0077); shown (by positive rays) to be free from isotopes, on the earth. 2. *Helium*, He (4·00); no isotopes. 3. *Lithium*, Li (6·94); mixture of isotopes Li_I (7) and Li_{II} (6). 4. *Glucinum*, Gl (9·1). 5. *Boron*, B (10·9); mixture of isotopes B_I (11) and B_{II} (10); loses H on bombardment with α particles. 6. *Carbon*, C (12·00); no isotopes. 7. *Nitrogen*, N. (14·01); no isotopes; loses H on bombardment with α particles. 8. *Oxygen*, (16·00); no isotopes. 9. *Fluorine*, F (19·00); no isotopes.; loses H on bombardment with α particles.

The following elements resemble helium and the elements that follow it, repeating their properties in the same order :—

10. *Neon*, Ne (20·2); mixture of isotopes Ne_I (20) and Ne_{II} (22). 11. *Sodium*, Na (23·0); loses H on bombardment with α particles. 12. *Magnesium*, Mg (24·3); mixture of isotopes Mg_I (24), Mg_{II} (25) and Mg_{III} (26). 13. *Aluminium*, Al (27·1); emits H on bombardment with α particles. 14. *Silicon*, Si (28·3); mixture of isotopes S_I (28) and S_{II} (29). *Phosphorus*, P (31·0); no isotopes; emits H on bombardment with α particles. 16. *Sulphur*, S (32·06). 17. *Chlorine*, Cl (35·456); mixture of isotopes Cl_I (35) and Cl_{II} (37). 18. *Argon*, A (39·9); mixture of A_I (40) and A_{II} (36). 19. *Potassium*, K (39·1); mixture of K_1 (39) and K_2 (41); emits β rays. 20. *Calcium*, Ca (40·07); mixture of Ca_I (40) and Ca_{II} (44). 21. *Scandium*, Sc (44·5).

Now follow ten elements of small atomic volume, without strictly analogous predecessors :—

22. *Titanium*, Ti (48·1). 23. *Vanadium*, V (51·0). 24. *Chromium*, Cr (52·0). 25. *Manganese*, Mn (55). 26. *Iron*, Fe (55·8). 27. *Cobalt*, Co (58·97). 28. *Nickel*, Ni (58·68); mixture of Ni_I (58) and Ni_{II} (60). 29. *Copper*, Cu (63·6).

---

[1] Where there are several isotopes they are arranged in order of decreasing importance.

30. *Zinc*, Zn (65·4); mixture of Zn_I (64) and Zn_{II} (70).
31. *Gallium*, Ga (70·1).

The analogy with the previous elements is now resumed, in the same order, at the point where the sequence was abandoned, with the element germanium :—
32. *Germanium*, Ge (72·5). 33. *Arsenic*, As (74·96); no isotopes. 34. *Selenium*, Se (79·2). 35. *Bromine*, Br (79·9); mixture in very nearly equal proportions of Br_I (79) and Br_{II} (81). 36. *Krypton*, Kr (82·9); mixture of six isotopes (84, 86, 82, 83, 80, 78). 37. *Rubidium*, Rb (85·45); mixture of Rb_I (85) and Rb_{II} (87); emits β rays, as K. 38. *Strontium*, Sr (87·63). 39. *Yttrium*, Y (88·7).

Now follows a second group of ten elements, analogous, element for element, to the first, and including a second atomic volume minimum :—
40. *Zirconium*, Zr (90·6). 41. *Niobium*, Nb (93·5). 42. *Molybdenum*, Mo (90·0). 43. Unknown. 44. *Ruthenium*, Ru (101·7). 45. *Rhodium*, Rh (102·9). 46. *Palladium*, Pd (106·7). 47. *Silver*, Ag (107·88). 48. *Cadmium*, Cd (112·4). 49. *Indium*, In (114·8).

We now find analogy with the elements that follow the first group of ten :—
50. *Tin*, Sn (118·7). 51. *Antimony*, Sb (120·1). 52. *Tellurium*, Te (127·5). 53. *Iodine*, I (126·92); no isotopes. 54. *Xenon*, X (130·2); mixture of at least 5 isotopes (128, 131, 130, 133, 135). 55. *Cæsium*, Cs (132·8); very feebly radioactive (β rays). 56. *Barium*, Ba (137·37).

Now come metals of the Rare Earth group, comparable to Sc or Y, but very analogous to each other :—
57. *Lanthanum*, La (139·0) 58. *Cerium*, Ce (140·2). 59. *Praseodymium*, Pr (140·6). 60. *Neodymium*, Nd (144·3). 61. Unknown. 62. *Samarium*, Sm (150·4). 63. *Europium*, Eu (152·0). 64. *Gadolinium*, Gd (157·3). 65. *Terbium*, Tb (159·2). 66. *Dysprosium*, Dy (162·5). 67. *Holmium*, Ho (163·5). 68. *Erbium*, Er (167·7). 69. *Thulium*, Tu (168·5). 70. *Neoytterbium*, Ny (172). 71. *Lutecium*, Lu (174).

A third group of ten elements, analogous to the earlier groups of ten :—
72. *Celtium*, Ct (known through its spectrum and its characteristic X-rays (Moseley's Law)). 73. *Tantalum*, Ta (181). 74. *Tungsten*, W (184). 75. Unknown. 76. *Osmium*, Os (191). 77. *Iridium*, Ir (193·1). 78. *Platinum*, Pt (195·0). 79. *Gold*, Au (197·2). 80. *Mercury*, Hg (200·5); mixture of at least four isotopes (197, 200, 202, 204). 81. *Thallium*, Th (204); and its short-lived isotopes *Thorium C'''* (208,

APPENDIX, 1921

period 3 minutes), *Actinium C″* (206, period 5 seconds), and *Radium C″* (210, period 1·4 seconds).   82. *Lead*, Pb (207), mixture of lead (206), the final element of the uranium series, and of lead (208), the terminal element of the thorium series. In addition there are 4 radioactive isotopes, Ra D (16 years), Th B (11 hours), Ac B (36 mins.), and Ra B (27 mins.).   83. *Bismuth*, Bi (208) and 4 radioactive isomers, Ra E (5 days), Th C (1 hour), Ra C (19 mins.), Ac C (2·2 mins.).   84. *Polonium*, Po (210); radioactive (period 136 days), emitting α particles and producing an element of atomic number 82 (*Lead* 206). Also Ra A (3 mins.) and 5 other short-lived isotopes, with periods less than 1 second (Ac C′, Th C′, Ra C′, Ac A, Th A).   85. Unknown halogen.   86. *Radium emanation* (*Radon*) (222) (3·85 days).   87. *Thorium emanation* (Thoron, Tn) (54 seconds); *actinium emanation* (Actinon, An) (218) (4 seconds).   87. Unknown alkali metal.   88. *Radium*, Ra (226·4); period 1800 years, and 3 isotopes, *Mesothorium*, Mth (228, period 5·5 years), *Actinium X* (11 days) and *Thorium X* (4 days).   89. *Actinium* Ac (226–227, period 30 years) and *Mesothorium*$_{II}$, Mth$_{II}$ (5 hours).   90. *Thorium*, Th (232·1, period $1\cdot 5 \times 10^{10}$ years) and 5 isotopes; *Ionium*, Io (230, period 10^5 years), *Radiothorium* (228, period 2 years), *Uranium* X_1 (period 24 years), *Radioactinium* (19 years), *Uranium Y* (1 year).   91. *Protoactinium*, Pa (10^4 years), and *Uranium* X_2 (1·15 mins.).   92. *Uranium*, U$_I$ (238·5, period 10^9 years), and *Uranium*, U$_{II}$ (234, period 10^6 years).

TABLE OF THE ELEMENTS.

(Atomic Numbers and Symbols.)

		1H	2He	3Li	4Gl	5B	6C	7N	8O								
		9F	10Ne	11Na	12Mg	13Al	14Si	15P	16S								
17Cl	18A	19K	20Ca	21Sc	22Ti	23V	24Cr	25Mn	26*Fe*	27*Co*	28*Ni*	29Cu	30Zn	31Ga	32Ge	33As	34Se
35Br	36Kr	37Rb	38Sr	39Y	40Zr	41Nb	42Mo	43?	44*Ru*	45*Rh*	46*Pd*	47Ag	48Cd	49In	50Sn	51Sb	52Te
53I	54X	55Cs	56Ba	57*/71	72Ct	73Ta	74W	75?	76*Os*	77*Ir*	78*Pt*	79Au	80Hg	81Tl	82Pb	83Bi	84Po
85?	86Rn	87?	88Ra	89Ac	90Th	91Pa	92U										

* Rare Earths.

57La 58Ce 59Pr 60Nd 61? 62Sm 63Eu 64Gd 65Tb 66Dy 67Ho 68Er 69Tu 70Ny 71Lu.

Mendéléjeff's Law is obvious, elements in the same vertical column being analogous.

INDEX

A.

AGITATION,
  molecular and diffusion, 4.
  and expansion of fluids, 6.
Arrhenius's hypothesis, 39.
Atom,
  gramme, 21.
  material part concentrated at centre, 161.
  of electricity, 173.
Atomic,
  coefficients, 19.
  disintegration, 197.
  hypothesis, 10.
  projectiles, visibility of paths, 214.
  weights, 21 et seq.
Atoms, 7.
  ageing of, 202.
  counting of, 207 et seq.
  dimensions of, 49—52
  relative weights of, 11, 12.
Avogadro's hypothesis, 17—19.
  proof of, 59.
Number, 26, 49.

B.

BLACK bodies, 144, 145.
  composition of light from, 148 et seq.
Bohr, 171, 220.
Boltzmann, 60 et seq.
Brownian movement, 83—89.
  and Carnot's principle, 86.
  definition of activity of, 110 et seq.
  in emulsions, 99.
  irregularity of, 116.
  rotational, 113, 124.

C.

CARNOT'S principle, 86.

Centrifuging, fractional, 94.
Charge,
  minimum elementary, 43, 48.
  on gramme ion, 44.
  on a projectile, 207.
Chemical,
  discontinuity, 9.
  formulæ, 13.
Constant, Planck's, 154.
Constitutional formulæ, 33.
Corpuscles, 186.
Cosmogeny, 204.
Crystals, liquid, 143.
  and X-rays, 175.

D.

DECOMPOSITION, 7.
Definite proportions, law of, 9.
Density, fluctuations of, 134.
Diffusion,
  of emulsions, 111.
  of large molecules, 127.
  of visible granules, 129.
Discontinuity,
  chemical, 9.
  of energy, 69—70.
Dissociation,
  electrolytic, 42.
  molecular, light as cause of, 163.
Distribution of grains in emulsions, 102.
Divisibility of matter, 48.
Dulong and Petit's law, 21.

E.

EFFUSION, 61.
Einstein's theory, 109.
  verification of, 114.
Electricity, atom of, 173.
Electrolytes, dissociation of, 42
Emanations, 199.

Emulsions, 89—106.
  gas laws, and, 89.
  preparation of, 94, 95.
Energy, discontinuity of, 69, 70, 73.
  quantum of, 70.
Equilibrium, in gas column, 90.
Equipartition of energy, 60.
Equivalents, 15.

F.

Films, thin, 49.
Fluctuations, 134—142.
Formulæ,
  constitutional, 33.
  molecular, 27.

G.

Gases,
  monatomic, 65.
  specific heat of, 69 *et seq.*
  viscosity of, 74.
Gas laws, extension to emulsions, 89.
Gay-Lussac, 18.
Gramme atom, 21.
  molecule, 26.

H.

Helium,
  number of atoms in known volume of, 211.
  radioactive production of, 200.
Hoff's, van't, law, 39.
Hypothesis,
  atomic, 10.
  Arrhenius's, 39.
  Avogadro's, 17, 19.
  Prout's, 24.

I.

Impact, molecular, 77—81.
Ionisation, 175.
Ions, 40.
  gaseous, charge on, 166—169, 179.
Isotopes, 24 *et seq.*, 218.

K.

Kathode rays, 173.

L.

Law,
  Bohr's, 171.
  of definite proportions, 9.
  Dulong and Petit's, 21.
  van't Hoff's, 39.
  of multiple proportions, 10
  Proust's, 9—11.
  Raoult's, 36.
  Stefan's, 147.
  Stokes', 97—99.
Light and quanta, 148.
Liquid crystals, 143.

M.

Magnetons, 193.
Matter, divisibility of, 48.
Masked transformations, 207.
Mean free path, 74.
Membranes, semi-permeable, 38.
Mendélejeff's rule, 24.
Mixtures, persistence of components in, 1.
Molecular
  agitation, 4.
  formulæ, 27.
  magnitudes, from Brownian movement, 122.
    determination of, 81, 107.
    from black body radiation, 155.
  orientation, fluctuations in, 143.
  size, upper limit of, 48.
  structure, 27.
Molecule,
  dislocation of, during reaction, 31.
  distribution of matter in, 162.
Molecules,
  critical, 167.
  diameter of impact of, 77.
  free paths of, 74.
  in constant impact, 71.
  rotation of, 67.
  size of, 48.

INDEX

Molecules—*continued.*
   velocities of, 53, 56.
   vibration of, 64.
Moseley's Law, 219.
Multiple proportions, law of, 9.

N.

NEBULIUM, 64, 223.
Nuclei, 221.
Number, Avogadro's, 49.
Numbers, proportional, 13.

O.

OPALESCENCE, critical, 135.
Osmotic pressure, 38.

P.

a PARTICLES, *see* a Projectiles.
Planck's constant, 154.
Positive rays, 180.
a Projectiles, 199—203.
Projectiles, atomic, 205.
Protection, sphere of, 66.
Proust's law, 9—11.
Prout's hypothesis, 24.
Pure substances, 2—3.

Q.

QUANTA, 152, 154, 171.
   and light, 147.
   and rotation, 159.
   and specific heats of solids, 154.

R.

RADIATION, 164 *et seq.*
Radioactive Equilibrium, 198.
   series, 194.
Radio-chemistry, 164.
Radioactivity, 195 *et seq.*
Radium, number of atoms in known mass of, 203.

Raoult's laws, 36.
a Rays, 200.
Rays,
   kathode, 164.
   positive, 189, 192.
   X-, 165.
Rotation, unstable, 159.

S.

SCINTILLATIONS, 207.
Semi-permeable membranes, 38.
Similar compounds, 15.
Simple substances, 7.
Sky, blueness of, 139—142.
Solution, 36.
Specific heat,
   gases, 69.
   solids, 22, 77, 156.
Spectral lines, width of, 62.
Sphere of protection, 66.
Stefan's law, 147.
Stereochemistry, 35.
Stokes' law, 97—99.
Substitution, 28.

T.

THIN films, 49.
Transmutation, 195.
   masked, 199.

V.

VALENCY, 31.
   bond, strength of, 161.
   electrical, 44.
Van der Waal's equation, 79.

X.

X-RAYS, 165.
  ,, characteristic, 219.
  ,, ultra, 224.